Mass Spectrometry

Mass Spectrometry

James McCullagh and
Neil Oldham

OXFORD
UNIVERSITY PRESS

UNIVERSITY PRESS

Great Clarendon Street, Oxford, OX2 6DP,
United Kingdom

Oxford University Press is a department of the University of Oxford.
It furthers the University's objective of excellence in research, scholarship,
and education by publishing worldwide. Oxford is a registered trade mark of
Oxford University Press in the UK and in certain other countries

Published in the United States of America by Oxford University Press
198 Madison Avenue, New York, NY 10016, United States of America

British Library Cataloguing in Publication Data

Data available

Library of Congress Control Number: 2019937519

ISBN 978-0-19-878904-8

Preface

Mass spectrometry is now one of the most important analytical techniques in the chemical and biochemical sciences, and its applications have expanded into many other fields of research. Rapid technical developments towards the second half of the twentieth century helped its transformation from an instrument with niche applications in the physics laboratory to one with extremely broad scientific utility. Modern mass spectrometry is applied in disciplines from archaeology to zoology, encompassing chemistry, biochemistry, plant science, space exploration, airport security, environmental monitoring, food authenticity, and a wide range of medical applications. It has been the subject of Nobel Prizes, both directly and indirectly, and continues to make significant contributions to scientific discovery.

In this book we aim to provide an accessible and concise introduction to the general principles and applications of modern mass spectrometry. Whilst the book is written primarily for undergraduate and graduate students studying chemistry and biochemistry courses, we hope that it will also serve as a point of entry and reference for those from a wider range of scientific backgrounds.

The chapters describing general principles cover methods of ionization, mass analysis, resolution, accurate mass, and sensitivity, as well as tandem mass spectrometry (MS/MS) and spectral interpretation. We have tried to show the diversity of uses of mass spectrometry and explain how the mass spectrometer is employed in these areas. The chapter on modern applications outlines a selected number covering environmental and medical analysis, metabolomics and proteomics, biophysics, structural biology, and imaging. Each of the chapters includes exercises and suggested further reading.

Contents

Introduction

1.1 What is mass spectrometry?

Mass spectrometry (MS) is an analytical technique that forms ions from atoms or molecules and measures their mass-to-charge ratios (m/z). It can provide information about molecular and elemental composition and also quantify the abundance of individual chemical components. It is a highly selective technique, meaning that it can differentiate between multiple compounds within a complex chemical or biological sample. It can be used to analyse solids, liquids, and gases with very high sensitivity (low picomole levels (approximately 1 trillionth of a gram) is typical for modern instruments). It is said that, for the amount of sample required, mass spectrometry can provide more structural information than almost any other technique, including crystallography and nuclear magnetic resonance spectroscopy (NMR).

A mass spectrometer measures the m/z of ions in the gas phase. This concise statement provides a number of clues about how the measurement process takes place and the type of analytical information that can be determined. First, the analysis deals with *ions* (charged atoms or molecules). Unlike neutral atoms and molecules, ions can be accelerated, deflected, and deaccelerated in electric and magnetic fields, which are crucial features for the operation of a mass spectrometer. Second, mass spectrometers measure *mass-to-charge ratio* and not simply *mass*; this is because the behaviour of an ion in an electric or magnetic field depends upon the number of charges (z) the ion possesses. Thus, in order to determine the mass of an ion by mass spectrometry, we must also determine z; fortunately this is usually straightforward to do using the mass spectrum (see Section 1.5). Third, analysis takes place in the *gas phase*. In order to achieve this, the sample needs to be converted into gaseous ions A high vacuum is required inside the mass spectrometer to allow gas-phase ions to pass through the instrument without colliding with molecules in the atmosphere.

To be amenable to analysis by mass spectrometry a molecule must be able to form a gas-phase ion that is stable for long enough to reach the detector of the mass spectrometer either intact or as fragment ions. For many years this was a

A mass spectrometer measures the mass-to-charge ratio (m/z) of ions in the gas phase.

In mass spectrometry, m refers to the mass of the ion in atomic mass units, which are given the symbol u or Da (daltons, after John Dalton). By definition, 1 u (or Da) = 1/12 mass of the carbon-12 atom (m (^{12}C)). z is the integer charge state of an ion (+1, +2, −1, −2, etc.) of total absolute charge ze, where e is the elementary charge in coulombs (1.60×10^{-19} C).

major challenge for the analysis of certain types of molecule, particularly large ones such as proteins, oligonucleotides, and synthetic polymers. An important part of the history of mass spectrometry has been the development of instrumentation capable of measuring compounds with a wider range of chemical and physical properties.

1.2 Atomic and molecular mass

1.2.1 Isotopically averaged mass, nominal mass, and monoisotopic mass

Isotopologues and isotopomers. Isotopologues are molecules that differ only in their isotopic composition. For example, $^{1}H_2O$, $^{1}H^2HO$, and $^{2}H_2O$ are all isotopologues. Isotopomers, in contrast, possess the same overall isotopic composition, but differ in the positions of those isotopes in the molecule. For example, $C^1H_3O^2H$ and $C^1H_2{}^2HO^1H$.

When referring to the atomic mass of a particular element it is common to use either the isotopically averaged mass of all of the naturally occurring isotopes, or the mass of a single isotope. The former, equivalent to the dimensionless relative atomic mass A_r, is the one shown on most periodic tables, and is widely employed to derive relative molecular mass (M_r) in many chemical calculations. Using ethanol (C_2H_6O) as an example, the relative atomic masses (to three decimal places) of the constituent elements are A_r (C) = 12.011, A_r (H) = 1.008, A_r (O) = 15.999, and thus the relative molecular mass of ethanol is 46.069.

For *small molecules* (< 900 Da) most modern mass spectrometers can easily resolve the 1 Da mass difference caused by the presence of the different isotopes of their constituent elements (for example ^{13}C versus ^{12}C). For this reason single isotope masses are normally used in mass spectrometry of small molecules rather than the isotopically averaged mass. The two commonly encountered types of isotopic mass are:

Small molecules are generally considered to be those with a molecular mass < 900 Da, although the cut-off is not precise. They include many organic molecules, drugs, metabolites, and metal complexes. They are generally contrasted by *large biomolecules* such as proteins or oligonucleotides and large synthetic polymers.

i) *Nominal mass*: mass of an ion or molecule calculated using the isotopic mass of the most abundant constituent elemental isotope rounded to the nearest integer value and multiplied by the number of atoms of each element.

ii) *Monoisotopic mass*: exact mass of an ion or molecule calculated using the atomic mass (m_a) of the most abundant isotope of each element multiplied by the number of atoms of each element.

In mass spectrometry, the difference between the nominal mass and monoisotopic mass is known as the *mass defect*.

Thus, for ethanol (C_2H_6O) the nominal mass of the molecule is 46 Da (m_a (C) = 12 Da, m_a (H) = 1 Da, m_a (O) = 16 Da), whilst the monoisotopic molecular mass is 46.043 Da (m_a (C) = 12.000 Da, m_a (H) = 1.008 Da, m_a (O) = 15.995 Da, to three decimal places).

As can be seen from the three different values for the molecular mass of ethanol, it is important to recognize, and for it to be stated clearly, which definition of mass is being used. For small molecules, where isotope separation is achieved but the mass measurement is accurate to 1 Da only, nominal mass is normally used. Where accurate mass measurements (> 3 decimal places) are made with isotopic resolution (see Chapter 4), then monoisotopic mass is used.

In cases where isotopic resolution is not achieved by the spectrometer, isotopically averaged mass is employed in order to ensure proper comparison between measured and calculated values.

1.2.2 **Mass is an intrinsic and differentiating component of matter**

The elements that comprise all chemical species have, by definition, a unique proton, neutron, and electron composition, and each sub-atomic particle has a unique, non-integer mass in the unbound state (see Table 1.1).

If nuclear mass, and thus atomic and molecular mass, were simply derived from the sum of the unbound masses of the appropriate number of protons and neutrons for the given elements in a compound, mass spectrometry would be limited in its ability to determine elemental composition. Whilst elements in the periodic table all have different proton numbers, the sub-atomic particle composition often reflects an exact multiple of other elements. Two oxygen (^{16}O) atoms, for example, have the same number of protons, neutrons, and electrons as a single sulfur (^{32}S) atom (see Table 1.2). Importantly, however, sub-atomic particles bound in the nucleus of an element do not have exactly the same mass as in the unbound state. A small amount of the mass of elemental nuclei is lost due to energy released during its formation (nuclear binding energy), and the amount released differs according to the size of the nucleus. Nuclear binding energy differences ensure that all elements, *and combinations of elements*, have a unique mass that is a non-integer value and not a multiple of other elements. The measurement of *m/z* with sufficient accuracy can therefore be used to determine a unique chemical formula. It is only relatively recently that mass spectrometers with high mass accuracy (three to five decimal places) have become widely available and this has made modern mass spectrometry an extremely powerful method for the determination of elemental composition. This subject is explored further in Chapter 4.

1.3 **What can mass spectrometry do?**

Before going on to look at the layout and components of a mass spectrometer, it is instructive to consider briefly what mass spectrometry can actually be used for. The most obvious function is in determining the molecular mass of an analyte (remember that mass spectrometry measures *m/z*, and *m* can be deduced providing *z* is known). As discussed previously, and at some length in Chapter 4, if this measurement can be made with high accuracy then it is possible to deduce the elemental composition of a small molecule.

As well as measuring molecular mass, mass spectrometry can also be used to generate fragment ions that provide structural information and act as a molecular 'fingerprint' for an analyte. For certain types of fragmentation, reproducible fragment ion ratios can be obtained that allow automated library searching and identification. Structural information is not necessarily inherent in the measurement of the *m/z* of an individual compound, but the *m/z* of its fragments can be used to interpret chemical structures. The high sensitivity and selectivity of mass spectrometry can make it the only practical tool for such structural analyses in some cases; for example, when an analyte is at very low abundance or has a large molecular weight.

Table 1.1 The mass of sub-atomic particles in the unbound state.

Sub-atomic particle	Mass (Da) (unbound state)
Proton	1.0073
Neutron	1.0087
Electron	0.0005

Definition of nuclear binding energy: The effective mass of the nucleus of an element (protons + neutrons) is less than the mass of its constituent sub-atomic particles in their free state. The mass difference is equivalent to the nuclear binding energy that is released during formation of the element. The magnitude of the nuclear binding energy is specific to the mass of each nucleus.

Table 1.2 The mass of selected elements in the bound and unbound state.

Mass (Da)	di-oxygen ($^{16}O_2$)	Sulfur (^{32}S)
Free state	32.2640	32.2640
Bound state	31.9898	31.9720

Mass (Da)	di-nitrogen ($^{14}N_2$)	Silicon (^{28}Si)
Free state	28.2317	28.2317
Bound state	28.0061	27.9769

As discussed previously and in Section 1.5, mass spectrometry is able to separate isotopes and hence provide information about the isotopic composition of molecules in a sample. In a specialist area, known as isotope ratio mass spectrometry, this information can be used to deduce the environmental and metabolic origin of a given compound. This is used, for example, to understand ecological food-webs and to reconstruct dietary information from the analysis of proteins preserved in archaeological bone.

Mass spectrometers are also able to act as sensitive detectors, especially when coupled to a chromatograph (see Chapter 7). The combination of mass spectrometry with either gas or liquid chromatography (GC or LC) provides an extremely powerful platform for the analysis of trace components within biological, pharmaceutical, or environmental samples (see Chapters 7 and 8). This permits the analysis of illicit drugs, tracing impurities, and identifying pollutants in the environment, for example. The coupling of chromatography with mass spectrometry also provides a platform for the quantification of chemical species which can be difficult to achieve by mass spectrometry alone.

A significant application of mass spectrometry in recent years has been in protein identification and sequencing (see Chapters 6 and 8). The ability of mass spectrometry to provide this information with high sensitivity in complex samples has transformed the field of proteomics (the study of the protein composition of an organism, system, or biological context).

1.4 The mass spectrometer

1.4.1 The use of electric and magnetic fields to manipulate ions

All mass spectrometers use electric or magnetic fields to accelerate and/or manipulate ions to measure their m/z. The relationship between the mass, charge, and force applied by an electromagnetic field is described principally by two basic equations:

$F = ma$ (Newton's second law)

$F = ze(E + v \times B)$ (Lorentz force law)

where

F is the force applied to the ion by the electromagnetic field

m is the mass of the ion

a is the acceleration

ze is the charge of the ion

E is the electric field

$v \times B$ is the vector cross product of the ion velocity and applied magnetic field

In the context of ions, Newton's second law demonstrates that an applied electromagnetic force (F) causes an ion to accelerate in a way that is dependent on

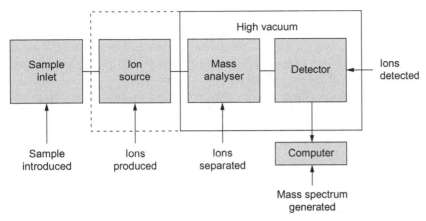

Figure 1.1 A schematic of the main parts of a generic mass spectrometer system.

mass, m. The Lorentz force equation goes on to show that this acceleration is also dependent on the charge of the ion, ze. This is why mass spectrometers measure the m/z of ions and not just their mass. It is these physical laws that govern how charged species, once formed in an ion source, can be manipulated and separated inside a mass spectrometer.

The schematic in Figure 1.1 represents the five basic components of a generic mass spectrometer. In general terms, a sample is introduced via a *sample inlet* and ions are formed from neutral molecules in an *ion source*. The ions are then accelerated out of the source by a potential difference and guided to a *mass analyser*, which separates ions into the different m/z values present. Following separation, ions strike a *detector*, which records their abundance, usually via amplification of the ion signal. This output is then digitized and transferred to a *computer*, where the spectrum (a plot of ion abundance versus m/z) is displayed.

1.4.2 Sample inlets

The process of mass analysis begins with the introduction of the sample via an inlet system. The inlet serves the important function of getting the sample into the mass spectrometer efficiently. There are many types of inlet from a simple vapour reservoir, or solids probe, to more complex gas or liquid chromatography interfaces which enable separation of compounds prior to mass spectrometric analysis.

1.4.3 Ionization sources

Many ionization sources are available for mass spectrometers, although six are routinely used for the majority of applications. These are electron ionization (EI), chemical ionization (CI), electrospray ionization (ESI), atmospheric chemical ionization (APCI), inductively coupled plasma ionization (ICP), and matrix assisted laser desorption ionization (MALDI) (see Chapter 2). They differ in the physical mechanism of ion formation and in their suitability for different types

$$M + e^-_{(fast)} \longrightarrow M^{+\bullet} + 2e^- \text{ (EI)}$$
$$M + H^+ \longrightarrow [M + H]^+ \text{ (ESI positive)}$$
$$M - H^+ \longrightarrow [M - H]^- \text{ (ESI negative)}$$

Figure 1.2 Formation of ions by EI and ESI.

of sample (based on polarity, volatility, and other chemical and physical properties). The ionic species formed are not always of the same type; for example, EI leads to the production of radical cations ($M^{+\bullet}$) from vaporized samples, whereas ESI forms protonated ($[M+H]^+$) or deprotonated molecules ($[M–H]^-$) and is used for the analysis of compounds from solution (Figure 1.2).

In general, protonated/deprotonated molecules or radical cations/anions are the most common charged species formed in mass spectrometers, but the mechanisms by which they are produced can differ significantly. Chapter 2 provides a detailed introduction to the main types of ionization and ion sources used in modern mass spectrometry.

The process of ionization leads to transfer of energy to the newly formed ion raising its internal energy, which may in turn lead to fragmentation. The degree of energy transfer is highly dependent on the type of ionization method used and the structure of the ion. Electron ionization, for example, results in significant energy transfer, and subsequent fragmentation of the radical cation is common; this is consequently often referred to as a 'hard' ionization technique. For other ionization methods, such as electrospray ionization and matrix assisted laser desorption ionization, the energy changes can be much smaller, and often below the threshold required to break bonds. These are often referred to as 'soft' ionization techniques.

1.4.4 **Mass analysers**

The mass analyser is at the heart of the mass spectrometer, and is designed to separate ions of different m/z. Different mass analyser designs (see Table 1.3) possess a range of performance characteristics. Some provide unit mass resolution and relatively modest mass accuracy (e.g. quadrupole mass analysers), whilst others exhibit extremely high resolving powers and are capable of high mass accuracies (e.g. Fourier transform-ion cyclotron resonance (FT-ICR) mass analysers). It is also common to find more than one mass analyser in a modern mass spectrometer. These 'tandem instruments', enable selection of single m/z values and controlled fragmentation for the purposes of structural analysis, or to improve signal-to-noise ratios. Mass analysers are discussed further in Chapter 3, and tandem mass spectrometry is described in Chapter 5.

Table 1.3 Types of mass analyser.

Mass analyser	Abbrev.	Date
Double focusing magnetic sector	Sector	1932
Time-of-flight	TOF	1946
Quadrupole	Q	1953
Ion trap	IT	1953
Ion-cyclotron resonance	FT-ICR	1974
Orbitrap	Orbi	2005

1.4.5 **Detectors**

Mass spectrometer detectors generate an amplified signal from ions that have been separated by the mass analyser. The amplitude of the signal produced is proportional to ion abundance. There are a number of different detector types, including *Faraday cups*, which detect a current produced from ions captured by a metal surface; *electron multipliers*, which generate an electron cascade that amplifies an electric current produced by the arrival of an ion at the detector; and *photomultipliers*, which produce a photon cascade in a manner analogous to the electron multiplier. The *microchannel plate* (MCP) detector is a variant on the electron multiplier and is made of a silica wafer with an array of microscopic channels that each produce an electron cascade upon the arrival of an ion.

1.5 The mass spectrum

The output of a mass spectrometer is known as a *mass spectrum* (plural *spectra*). This is a plot of relative ion abundance on the vertical axis versus *m/z* on the horizontal axis. Figure 1.3 shows an example of a mass spectrum. It was generated by electron ionization (EI), and therefore contains a radical cation arising from removal of an electron from the molecule (see Figure 1.2). This is known as the *molecular ion* (M$^{+\bullet}$). The spectrum also displays *fragment ions* caused by dissociation of M$^{+\bullet}$ in the EI source. The most abundant ion (*m/z* 105 in this example) is termed the *base peak*. Mass spectra produced by softer ionization techniques, such as electrospray ionization (ESI), exhibit much less fragmentation of the protonated or deprotonated molecule ([M+H]$^+$ or [M−H]$^-$), and these species are usually also the base peak.

1.5.1 Isotopic distributions

As mentioned in Section 1.2, for small molecules most mass analysers are easily able to separate the 1 Da mass difference caused by the presence of heavier isotopes of the same element. This allows the measurement of isotopic ratios within the sample molecule. Figure 1.4 shows a close-up of the molecular ion

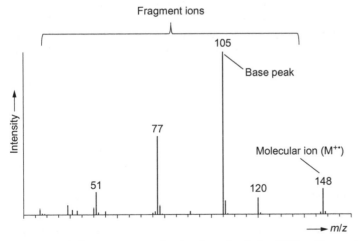

Figure 1.3 An example EI-mass spectrum showing the molecular ion (M$^{+\bullet}$) and fragment ions. The most abundant ion is known as the base peak.

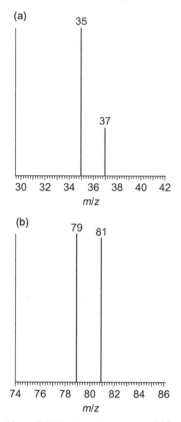

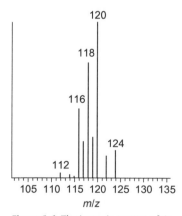

Figure 1.5 The isotopic patterns of (a) chlorine and (b) bromine.

Figure 1.6 The isotopic pattern of tin.

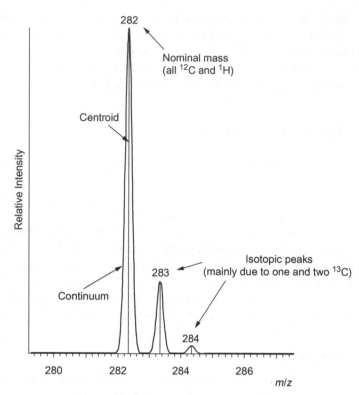

Figure 1.4 The molecular ion region of the mass spectrum of eicosane, $C_{20}H_{42}$, shown as continuum and centroid plots.

of the hydrocarbon eicosane, $C_{20}H_{42}$. The nominal mass, 282, is indicated, as are the isotopic peaks at 283 and 284, which arise due to the natural presence of heavier isotopes of carbon and hydrogen. The principal contribution to the isotopic peaks is from ^{13}C, which has a natural abundance of 1.07%. Deuterium (^{2}H) constitutes only 0.01% of hydrogen, and thus is only a very minor contributor to the signals at 283 and 284. The natural abundances of ^{14}C and ^{3}H are far too low to detect, and so the ion at 284 arises principally due to the presence of two ^{13}C atoms in the molecule.

Molecules containing additional elements exhibit other isotope patterns. For example, after ^{16}O (99.757%) the next most abundant isotope of oxygen is ^{18}O (0.205%), which means that oxygen makes a greater contribution to the second isotopic peak (+2 Da) than the first (+1 Da). The effect is even more pronounced with sulfur (^{32}S 94.93%, ^{34}S 4.29%). The halogens Cl and Br produce very distinctive isotopic patterns due to $^{35}Cl/^{37}Cl$ and $^{79}Br/^{81}Br$ (see Figure 1.5), which makes identifying the presence of these elements in molecules relatively straightforward. Likewise, some metals, such as chromium, iron, nickel, copper, zinc, and tin, have characteristic and often complex isotopic distributions. Tin, for example, has ten naturally occurring stable isotopes, all at appreciable concentrations (see Figure 1.6).

1.5.2 Continuum and centroid spectra

A mass spectrum is acquired by the instrument's detector as a continuous analogue response across the *m/z* scale. This signal is digitized for handling by a computer, which is able to reconstruct a continuum display by joining the discrete data points across the peak with a smoothed line. The result, known as a *continuum spectrum*, is shown in Figure 1.4. The advantage of this display is that it reveals the resolution achieved in the measurement: in other words, how well neighbouring *m/z* peaks are separated. The disadvantage is that data files are often large (megabytes of data) due to the high number of data points collected. An alternative way to display (and save) the data is as a *centroid spectrum* (also seen in Figure 1.4). This shows only the centre of each peak as a 'stick' plot, with one *m/z* and one intensity value. The centroid spectrum is often used for small molecules where unit mass resolution is achieved and nominal mass accuracy reported. In cases where it is necessary to show how well *m/z* peaks are resolved, a continuum spectrum is required.

1.5.3 Charge state

So far we have only considered mass spectra where the ions are singly charged, $z = 1$. This is almost always the case for small molecules ionized by EI. Larger molecules, such as peptides and proteins, usually exhibit higher charge states when ionized by electrospray ionization. This raises the question of how the charge state of an ion can be determined. If the mass of the molecule is known then the problem is a trivial one, as the charge will be that fraction of the mass at which the ion appears on the *m/z* scale. For example, a peptide of mass 2085.07 Da will give rise to a singly-protonated ion ($[M+H]^+$) at *m/z* 2086.07, whereas the doubly-protonated ion ($[M+2H]^{2+}$) will be at *m/z* 1043.54, and the triply-protonated ion ($[M+3H]^{3+}$) at *m/z* 696.03 (note the need to take into account the mass of each added proton).

If the molecular mass of the analyte is unknown, then we cannot use this information to determine the charge state. Providing that some degree of isotopic resolution is seen in the spectrum, then the separation of peaks in the isotopic cluster can be utilized to identify the charge state without any prior knowledge of molecular mass. This is because we know that the mass separation between neighbouring isotopic peaks ($\Delta m/z$) is 1 Da (the mass of the neutron). When the charge state $z = 1$ on the *m/z* scale, isotopic peaks will be separated by 1 (1/1). When the charge state $z = 2$, isotopic peaks will instead be separated by $\Delta m/z = 0.5$ (1/2), and for $z = 3$ the *m/z* separation between isotopic peaks will be $\Delta m/z = 0.33$ (1/3). This is illustrated in Figure 1.7, which shows the $[M+H]^+$, $[M+2H]^{2+}$, and $[M+3H]^{3+}$ for the hypothetical peptide of mass 2085.07 discussed previously. Since we know the *m/z* values from the measurement and can identify the charge state from isotopic separation in the spectrum, we can determine the mass of multiply-charged ions.

As can be seen in Figure 1.7, separation of isotopes on the *m/z* scale becomes less and less pronounced as *z* increases. Once the isotopes are no longer

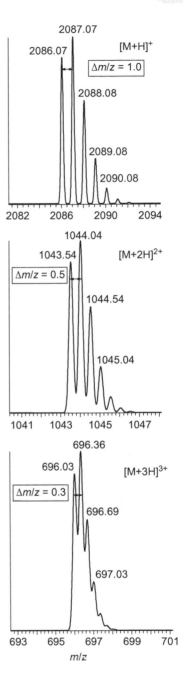

Figure 1.7 The singly-, doubly-, and triply-protonated molecules of a hypothetical peptide of mass 2085.07 Da, showing how isotopic separation can be used to deduce each charge state.

resolved, then it is impossible to assign the charge state using this method, and other approaches are required. This is one reason why mass analysers with high resolving powers are particularly advantageous for larger biomolecules. The different mass analysers used in mass spectrometry are discussed in Chapter 3, and the topic of resolution is covered in some detail in Chapter 4.

1.6 A brief history of mass spectrometry

In the latter part of the nineteenth century, before the discovery of sub-atomic particles, a number of scientists in Europe became interested in the effects of electrical discharges in purified gases. Improved vacuum technologies enabled these experiments to be conducted at low pressure, which led to the discovery that both positively and negatively charged 'rays' were formed, and that these could be manipulated by electric and magnetic fields. In 1887, Joseph John Thomson, a researcher at the Cavendish Laboratory in Cambridge (UK), measured the m/z ratio of negatively charged particles and went on to demonstrate the existence of the electron, for which he won the Nobel Prize for Physics in 1906. Soon afterwards, Eugene Goldstein was able to show that positively charged particles could be produced in a similar apparatus and that these 'canal rays', as they were known, were deflected by (stronger) electromagnetic fields. Goldstein was the first to measure the m/z ratio of the H^+ ion. J. J. Thomson and his research assistant Francis William Aston then built an improved instrument consisting of an electric discharge vessel containing an analyte gas at low pressure. After an electrical discharge produced positively charged particles, these ions were then extracted into a cathode tube and deflected by a parallel electric and magnetic field. The instrument was able to focus the resulting beam onto a flat fluorescent screen which measured their parabolic trajectory. Thomson and Aston had built the first mass spectrometer which generated ions, extracted and deflected them, and recorded their angle of deflection on a detector plate.

Discovery of isotopes: J. J. Thomson and F.W. Aston experimented with the formation of cations from atomic elements including mercury, argon, krypton, and neon. They found that neon always produced two ion beams calculated to be approximately mass 20 and 22. This was the first discovery that certain elements can exist in more than one isotopic form. Aston went on to show that many of the lighter elements possess multiple isotopes, for which he was awarded the Nobel Prize for Chemistry in 1922.

Improved mass spectrometers were developed in Europe and subsequently the USA in the 1930s, and in the 1940s the US petroleum industry started to use mass spectrometers to analyse oils. This led to the first commercial instruments which enabled an increasing number of laboratories to start using mass spectrometry. The second half of the twentieth century saw a dramatic increase in the use of mass spectrometers in both academia and industry. With the development of new ionization techniques towards the end of this period, mass spectrometry became an increasingly powerful tool in biology and medicine. Contemporary biological applications include sequencing proteins, analysis of DNA, and imaging biological tissues, and mass spectrometry now has applications in almost all branches of science. A timeline of significant technical developments and important applications of mass spectrometry is provided in Figures 1.8 and 1.9.

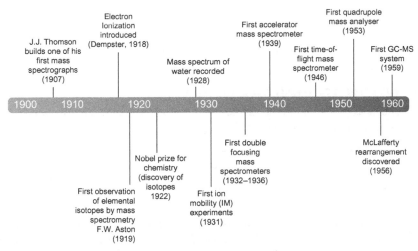

Figure 1.8 Mass spectrometry timeline 1900–1960.

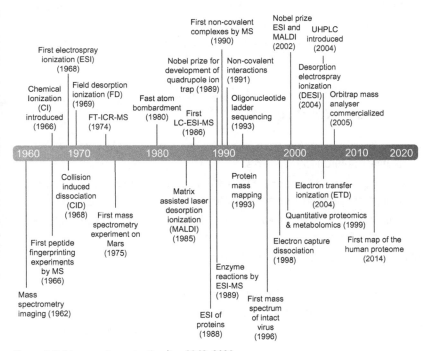

Figure 1.9 Mass spectrometry timeline 1960–2020.

1.7 Summary

From the material presented in this chapter you should be familiar with the following:

- What mass spectrometry is and what it can do.
- The different definitions of atomic and molecular mass, including average, nominal, and monoisotopic mass.
- How electric and magnetic fields can be used to deflect ions, and how the force applied to an ion is a function of both mass and charge.
- The basic layout of a generic mass spectrometer and the functions of its constituent parts.
- The main features of a mass spectrum, including knowledge of terms such as molecular ion, protonated/deprotonated molecule, fragment ion, base peak, continuum, and centroid spectra.
- The presence of and interpretation of isotopic patterns within a mass spectrum and how they can be used to provide information on elemental composition and charge state of the ion.
- The origins of mass spectrometry and some of the key developments in the field.

1.8 Further reading

Grayson, M. A. (ed.) (2002). *Measuring Mass: From Positive Rays to Proteins*. Philadelphia: Chemical Heritage Press.

Hoffmann, E. de and Stroobant, V. (2007). *Mass Spectrometry Principles and Applications*, 3rd edn. Chichester: Wiley.

Jennings, K. R. (ed.) (2012). *A History of European Mass Spectrometry*. Chichester: IM Publications.

Mallet, A. I. and Down, S. (2010). *Dictionary of Mass Spectrometry*. Chichester: Wiley.

Murray, K. K. et al. (2013). 'Definitions of terms relating to mass spectrometry (IUPAC Recommendations 2013)', *Pure Appl. Chem.* 85, 1515–609.

Watson, J. T. and Sparkman, O. D. (2007). *Introduction to Mass Spectrometry*, 4th edn. Chichester: Wiley.

2 Methods of ionization

2.1 Introduction

Ionization in mass spectrometry is the process of forming positively or negatively charged ions from analyte molecules. Ions can then be electrically or magnetically manipulated inside the high vacuum of a mass spectrometer to facilitate mass measurement. Ionization is therefore a fundamental process which ensures individual analytes can be separated by the mass analyser and detected. Indeed, the development of mass spectrometry as a pre-eminent analytical technique, with applications in modern physical, biological, and medical sciences, has been driven in large part by the development of new ionization methods.

A large number of ionization sources are available for modern mass spectrometers. This reflects the need for high sensitivity and extensive coverage of a very wide range of analytes. No technique is optimal for ionizing all analyte types, but many have overlapping capabilities. Which ionization method is most suitable for a particular compound depends on a number of factors, including chemical structure of the analyte, its polarity, stability, solubility, and whether it is solid, liquid, or gas at room temperature.

Ionization methods can be grouped according to the physical characteristics of the ionization process as follows:

1) Ionization of vaporized samples.
2) Desorption ionization.
3) Atmospheric pressure ionization.
4) Hybrid ambient ionization.

Ionization sources are almost always located immediately after, or combined with, the mass spectrometer inlet (where a sample is introduced into the instrument), either inside or outside the high vacuum of the system. Irrespective of the type of ionization method, or its location, an important function is the formation of a maximal number of ions, from neutral species of interest, and their efficient transfer into the mass spectrometer. The choice of ionization method for a particular experiment, however, is not always based solely on optimal sensitivity; it can also depend on factors such as how much

Ionization processes can differ significantly in the amount of energy transferred to the analyte during the ionization process, and this leads to two main types:

1) Those where energy transfer is relatively high leading to fragmentation of the analyte ion, known as 'hard' ionization techniques.

2) Those which transfer relatively little energy, and lead to minimal or no fragmentation, known as 'soft' ionization techniques.

sample is available, its physical properties, and whether quantitative or qualitative results are required.

In this chapter we will examine the most common ionization techniques in mass spectrometry. We will discuss different mechanisms of ion formation, basic ion-source designs, their relative performances, and areas of application.

2.2 Mechanisms of ion formation

Before going on to discuss the individual ionization techniques themselves we will first look at the general underlying processes that lead to ion formation. These can be characterized by a relatively small number of fundamental mechanisms, but it is worth bearing in mind that these can occur under quite different conditions. They may vary for different types of molecule, take place at different pressures and energies in different ion sources, and may be primary or secondary processes depending on the conditions.

The common ways in which analyte molecules become charged can be summarized as i) the removal or capture of an electron, ii) the removal or capture of a proton, iii) adduct formation with positive or negative ions, and iv) the formation of charged dimers or clusters. Figure 2.1 summarizes these mechanisms in schematic form.

Figure 2.1a shows a schematic of the formation of a radical cation. The mechanism by which this occurs can differ significantly depending on the type of ion source. For example, removal of an electron can be a high- or low-energy process depending on ion source conditions (contrast EI and FI in Sections 2.3.1.1 and 2.3.4, for example). Similarly, electron capture can also take place in a variety of ion sources where an electron interacts with an analyte molecule to form a negatively charged radical anion or (non-radical) anion. Here a number of mechanisms can occur, and sometimes do so simultaneously (Figure 2.1b–d). Electron capture is often exploited in the analysis of electrophiles, for example, using APCI and laser desorption ionization techniques (see Section 2.4).

Several ionization sources, such as electrospray ionization, MALDI, and a range of ambient ionization techniques (Section 2.4) exploit the addition of a proton or other cation (Na^+, K^+, for example), forming a positively charged ion–molecule pair in the gas phase (a protonated or cationated molecule). Figure 2.1e shows a general mechanism for the formation of a protonated molecule. In negative ion mode, removal of a proton or addition of an anion (Cl^- or NH_4^+ or OAc^-, for example) forms negatively charged molecules or adducts in the gas phase (Figure 2.1f). These processes of charged molecule formation in both positive and negative ion mode can be exploited under a variety of different ionization conditions in different types of ion source. It is particularly notable that a wide range of ion–molecule reactions are common during ionization at atmospheric pressure. This is due to the high probability of ion–molecule interactions when a large number of neutral gas molecules

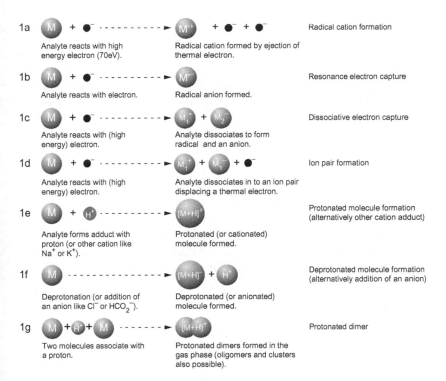

1a Analyte reacts with high energy electron (70eV). → Radical cation formed by ejection of thermal electron. Radical cation formation

1b Analyte reacts with electron. → Radical anion formed. Resonance electron capture

1c Analyte reacts with (high energy) electron. → Analyte dissociates to form radical and an anion. Dissociative electron capture

1d Analyte reacts with (high energy) electron. → Analyte dissociates in to an ion pair displacing a thermal electron. Ion pair formation

1e Analyte forms adduct with proton (or other cation like Na^+ or K^+). → Protonated (or cationated) molecule formed. Protonated molecule formation (alternatively other cation adduct)

1f Deprotonation (or addition of an anion like Cl^- or HCO_2^-). → Deprotonated (or anionated) molecule formed. Deprotonated molecule formation (alternatively addition of an anion)

1g Two molecules associate with a proton. → Protonated dimers formed in the gas phase (oligomers and clusters also possible). Protonated dimer

Figure 2.1 Common ionization mechanisms in mass spectrometry.

are present. Proton transfer also commonly occurs during MALDI (Section 2.4.1) where a matrix ion proton transfers to a neutral analyte molecule. In other ionization mechanisms, a proton, or another ion, can form an adduct during the phase transition, such as during evaporation of solvent droplets in ESI. Figure 2.1g illustrates the formation of a protonated dimer, in essence an extension of protonation described previously. Protonated dimers and similar aggregates can occur between two or more analyte molecules and a proton, and are often seen in ESI at higher analyte concentrations. The corresponding mechanism occurs in negative ion mode.

The remainder of this chapter will focus on introducing the common ionization sources used in mass spectrometry. We will discuss the principles behind the ionization processes, their specific capabilities, and their areas of application.

2.3 Techniques for the ionization of vaporized samples

There are a number of ionization sources that require analytes to be vaporized under vacuum. These make up the first group to be discussed, and include some of the first ionization processes to be used in mass spectrometry.

2.3.1 **Electron ionization (EI)**

Electron ionization (EI) (previously called 'electron impact') was first described by Arthur J. Dempster in 1918. It was used almost exclusively in mass spectrometry until the late 1960s when other ionization sources were developed, and is still commonly used today. It is efficient for the ionization of small, thermally stable, organic compounds in the gas phase. The ion formation process utilizes a high-energy electron beam to produce molecular ions from neutral gas-phase molecules by removing an electron. Transfer of excess energy during ion formation often leads to rapid dissociation of the newly formed molecular ion, creating charged and uncharged fragments (see Chapter 5). The charged fragments often provide a unique spectral 'fragmentation fingerprint' that is highly reproducible. Modern EI-MS systems use automated online library searching of fragment spectra for routine identification of compounds. Fragmentation patterns can also be used for the interpretation of elemental and structural information, particularly useful for the analysis of novel compounds (see Chapter 6).

'Electron ionization' used to be known as 'electron impact ionization'. This name has fallen out of favour due to the implication that electrons actually impact neutral molecules. This is not the case due to the very large electrostatic repulsion between electrons and their very small size making any such collision very unlikely. The ionization process instead occurs due to large fluctuations in the electric field around the molecule caused by the close proximity of high energy electrons. This results in a rapid increase in energy of the molecule and subsequent electron ejection.

2.3.1.1 **EI: principles of ion formation**

The process of molecular ion formation occurs rapidly, initiated by the interaction between a high-energy electron (typically 70 eV) and a neutral analyte molecule. Rapid transfer of energy, from the high-energy electron to the neutral molecule, manifests in the promotion of a number of electrons into excited states. The internal energy of the neutral molecule increases, and on average 14 eV is transferred from a high-energy electron at 70 eV. Figure 2.2 shows a schematic of this ionization process.

Redistribution of internal energy is typically fast, and some excited states revert quickly back to lower-energy states via photon emission. The most loosely bound electron (typically non-bonding electrons on heteroatoms, if present) may be completely removed, which leads to ionization by formation of a radical cation. The location of charge, and which bonds break, is a subject of focus in

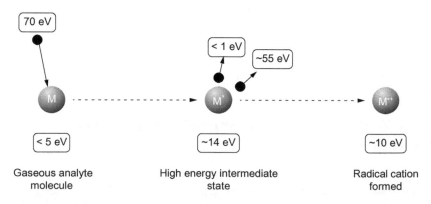

Figure 2.2 During electron ionization a high-energy electron displaces a molecular electron to produce a radical cation.

Chapter 6, but it is worth noting here that the fragments produced are often pre-dictable via relatively straightforward ion decomposition pathways which can provide useful structural information.

2.3.1.2 EI: source design and function

The electron ionization source is located on the high vacuum side of the mass spectrometer, adjacent to a heated sample inlet which delivers sample molecules into the ion source in the gas phase. The source is comprised of a reaction cham-ber with an external tungsten or rubidium filament which is used to generate high-energy electrons via thermionic emission. Once generated, the electrons are accelerated by an offset voltage (typically 70 eV) into the reaction chamber through a focusing slit. The continuous electron beam traverses a 15–20 mm gap where ionization of neutral analyte molecules takes place (see Figure 2.3). A superimposed magnetic field leads to a tightly controlled helical electron trajec-tory. This electron motion increases the probability of interaction with analyte gas molecules. An anode trap is located opposite the electron entrance slit to capture the high-energy electrons beyond the ionization space. Orthogonal to the electron beam lies a plate with a focusing slit which allows newly formed analyte ions to exit the source chamber towards the mass analyser. Opposite this exit lies a repeller plate which maintains a positive charge to push newly formed cations towards the exit plate. Just beyond this are a number of lenses which focus the newly formed ion beam and provide an acceleration potential towards the mass analyser.

2.3.1.3 EI: performance and applications

EI has stood the test of time. It was developed early in mass spectrometry, and remains in use for ionization of most organic compounds that are amenable to

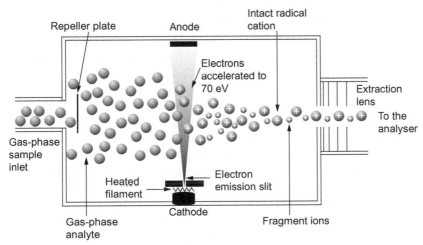

Figure 2.3 A schematic showing the layout of an electron ionization (EI) source and formation of positively charged radical cations.

gas-phase analysis. It also provides one of the most quantitative and reproducible ionization processes in mass spectrometry. It is, however, a low-efficiency ionization technique which, coupled with the abundant dissociation of molecular ions, leads to limits of detection in the picomolar range which is modest compared to other ionization techniques. The need to vaporize samples prior to EI also limits applications to the analysis of volatile, thermally stable compounds, and in practice it is applied only to analytes with a molecular weight < 1000 Da.

The simplicity and ruggedness of the EI source, coupled with the need for a gas interface, means that it is often found in instrument configurations involving gas chromatography and quadrupole or time-of-flight mass analysers (see Chapters 3 and 7). A recent update to these configurations is the GC-EI-orbitrap, which combines an EI ion source with the ultra-high-resolution, sensitivity, and accurate mass stability of the orbitrap mass analyser. This has recently found application in metabolomics and environmental residue analysis amongst others. Despite some drawbacks, EI remains an important ionization technique for chemical, pharmaceutical, forensic, and some biochemical applications.

2.3.2 Chemical ionization (CI)

The chemical ionization (CI) source was first developed in 1966 by Munsen and Field and quickly became important for the analysis of compounds which were too unstable for EI. It shares a similar design to EI and relies on radical cation formation to initiate the ionization process; however, a 'reagent gas' is ionized initially rather than the analyte itself. CI spectra are characterized by very little fragmentation and the formation of protonated or deprotonated molecules rather than radical cations or anions. Its applications are therefore often complementary to EI.

2.3.2.1 CI: principles of ion formation

The mechanism of CI starts with an electron beam ionizing reagent gas molecules via a classical EI process. This leads to the formation of reagent-gas radical cations which initiate a series of further ionization processes in the gas phase. Methane and ammonia are common reagent gases, but others can also be used and tailored to most efficiently ionize a particular analyte of interest. The CI ion source pressure is relatively high compared to EI, and the reagent gas pressure is significantly higher than the analyte pressure. Reagent-gas radical cations are therefore formed preferentially but rapidly interact with neutral reagent-gas molecules. This process leads to a cascade of ion–molecule reactions which ultimately manifest in proton transfer to the neutral analyte molecule. Figure 2.4 shows a schematic of a typical ion formation process and Figure 2.5 shows a typical cascade of ion–molecule and ion–ion reactions that can occur during the CI process using methane as a reagent gas. Table 2.1 summarizes the terminology used to refer to the different forms of ions generated by mass spectrometers.

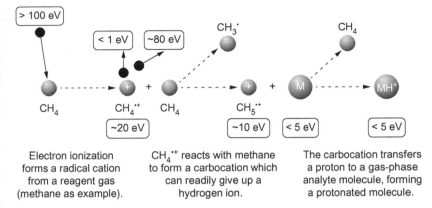

Electron ionization forms a radical cation from a reagent gas (methane as example).

$CH_4^{\bullet+}$ reacts with methane to form a carbocation which can readily give up a hydrogen ion.

The carbocation transfers a proton to a gas-phase analyte molecule, forming a protonated molecule.

Figure 2.4 A schematic showing ion formation during chemical ionization (CI).

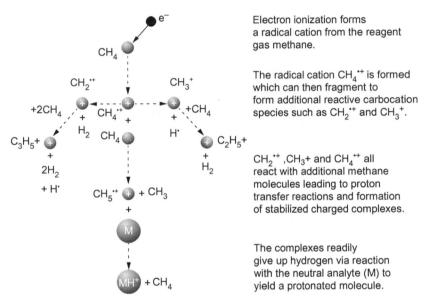

Electron ionization forms a radical cation from the reagent gas methane.

The radical cation $CH_4^{\bullet+}$ is formed which can then fragment to form additional reactive carbocation species such as $CH_2^{\bullet+}$ and CH_3^+.

$CH_2^{\bullet+}$, CH_3+ and $CH_4^{\bullet+}$ all react with additional methane molecules leading to proton transfer reactions and formation of stabilized charged complexes.

The complexes readily give up hydrogen via reaction with the neutral analyte (M) to yield a protonated molecule.

Figure 2.5 The detailed ion–molecule and ion–ion reactions associated with the formation of protonated analyte molecules using methane as the reagent gas during chemical ionization (CI).

Different reagent gases provide the ability to select optimal conditions for charge transfer, and the presence of protonated molecular ions implies that the analyte has a proton affinity that is at least comparable to that of the reagent gas (Table 2.2).

The CI process can often produce negatively charged ions with very high efficiency (unlike EI). It is interesting to note that the mechanism for doing so is very different from the formation of positive ions by CI. 'Electron capture' is involved,

Proton affinity (PA) is defined as the negative value of the enthalpy change in the reaction between a species (anion, molecule, or atom) and a proton in the gas phase, as illustrated for methane:

$$CH_4 + H^+ \rightarrow CH_5^+ \ [552 \text{ kJ mol}^{-1}]$$

Table 2.1 Terminology used for referring to the charged species formed during ionization in mass spectrometry.

Species formed during ionization	Correct terminology	Incorrect terminology
$M^{+\cdot}$	Radical cation or molecular ion	–
$M^{-\cdot}$	Radical anion or molecular ion	–
$[M+H]^+$	Protonated molecule	Protonated molecular ion Pseudo-molecular ion Quasi-molecular ion Molecular ion
$[M-H]^-$	Deprotonated molecule	Deprotonated molecular ion Pseudo-molecular ion Quasi-molecular ion Molecular ion
$[M+Na]^+$	Sodiated molecule	Pseudo-molecular ion Quasi-molecular ion Molecular ion
$[M+K]^+$	Potassiated molecule	Pseudo-molecular ion Quasi-molecular ion Molecular ion

Table 2.2 Common reagent gases used in chemical ionization (CI) and their proton affinities.

Reagent gas	Protonated/ionized species	Proton affinity (kJ mol^{-1})
H_2	H_2^+	424
CH_4	CH_5^+	552
H_2O	H_3O^+	697
NH_3	NH_4^+	854
CH_3OH	$CH_3OH_2^+$	762
NO	NO^+	531
O_2	O_2^+	422
CO	CO^+	594

and the process may proceed via a number of different routes depending on the energy of the electron. Figure 2.6 illustrates some common mechanisms, and typical energies by which electron capture occurs in CI. Some fragments can be observed via dissociative electron capture and ion-pair production, often as a result of interactions with higher-energy electrons.

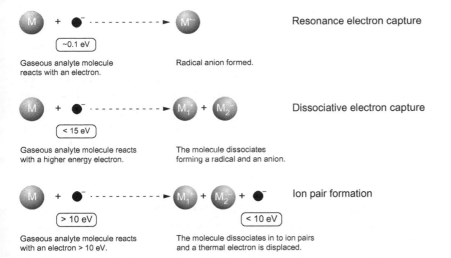

Figure 2.6 Types of electron capture reactions associated with negative ion formation during chemical ionization (CI).

2.3.2.2 CI: source design and function

The ion source design is similar to EI but with a number of important differences. The dimensions of the ionization chamber are smaller and the exit slit narrower so that a higher local pressure of reagent gas can be achieved. Although the pressure in this region must be relatively high, it must not compromise maintaining a lower pressure in the ion extraction region and into the rest of the mass spectrometer. Differential pumping systems in CI instruments help maintain the relative pressure differences. Electron generation is via a metal wire filament as in EI, but usually much higher electron energies are generated in the region of several hundred eV to ensure that electrons can penetrate the region of higher reagent-gas pressure. Figure 2.7 provides a schematic to illustrate the CI source design.

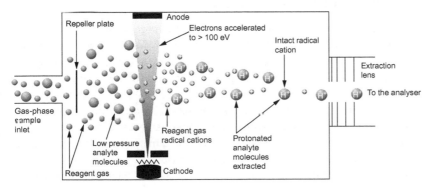

Figure 2.7 A schematic showing the layout of a chemical ionization (CI) source and the ion formation of protonated molecules.

2.3.2.3 CI: performance and applications

Many small organic compounds, including some amines, alcohols, nitriles, and nitro-containing compounds, readily decompose via EI and often do not produce an observable molecular ion. For these, and other compounds too unstable to form a molecular ion directly, CI is often a more useful ionization technique. The possibility for negative ion formation also enables more acidic compounds to be analysed that are also not amenable to the EI process. Being a 'soft' ionization technique, the protonated or deprotonated molecules exhibit little fragmentation, and so CI can be significantly more sensitive than EI. Similar to EI, it requires molecules to be in the gas phase and can be linked directly to a gas chromatography inlet (Chapter 7). In contrast to EI it provides almost no structural information, and subsequently there are no fragmentation libraries for compound identification. Despite its complementary performance, CI is also only amenable to volatile compounds that are stable enough to remain intact in the gas phase, and it is therefore also limited to analysis of compounds < 1000 Da in mass.

2.3.3 Inductively coupled plasma ionization (ICP)

Inductively coupled plasma (ionization) mass spectrometry (ICP-MS) was first commercialized in 1983. It uses plasma to generate cations from a gaseous, nebulized liquid or volatilized solid sample, with the latter often produced by laser ablation. By using hot plasma it can convert molecules into free elemental ions. ICP is able to ionize most of the elements in the periodic table that readily form positively charged ions, and is therefore particularly useful for the analysis of metals down to trace levels. It is extremely sensitive, routinely achieving parts per trillion detection limits, and has one of the highest linear responses to concentration of all the ionization techniques used in mass spectrometry. In addition to straightforward elemental analysis, ICP-MS is able to measure individual isotope abundances of the elements very accurately.

2.3.3.1 ICP: principles of ion formation

The mechanism of ion formation is based on the process of electron removal from analyte atoms. The sample, in gaseous or aerosol form, is introduced into an argon plasma. This is a mixture of argon atoms, ions, and electrons at temperatures typically > 6000 °C. Rapid drying and atomization of the sample is followed by absorption of energy, leading to atomic ionization. A^+ and A^{2+} atomic ions are most commonly formed, and these are then selected via a sampler cone, cooled in an intermediate vacuum chamber, and subsequently extracted into the mass spectrometer via a skimmer cone. Quadrupole and magnetic sector type analysers are commonly used for m/z measurements (see Chapter 3 for discussion of mass analysers). Figure 2.8 provides a schematic of the ICP ion source design.

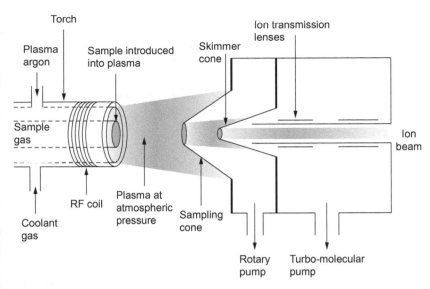

Figure 2.8 A schematic showing the generic layout of an inductively coupled plasma-mass spectrometry (ICP-MS) ion source.

A feature of ICP-MS is that all molecular information is lost through atomization in the ion source, but this highly efficient ionization process leads to a very sensitive and quantitative response to elemental composition.

2.3.3.2 ICP: performance and applications

ICP-MS has been interfaced with gas and liquid chromatography (GC-MS and LC-MS) since the late 1980s, and applications are extensive. These are found particularly in geochemistry, geochronology, and archaeology where isotopic analyses can provide chronological information on the timescales associated with geological evolution. Environmental monitoring and radioactive waste control are also important applications that benefit from the technique's high sensitivity. Areas of more recent application include quantitative proteomics in which sulfur and phosphorous analysis have been used to quantify absolute protein concentrations, and the use of ICP-MS as a quantitative detector for immuno-assays and immuno-imaging is currently an area of developing interest.

2.3.4 Field ionization (FI)

FI was developed by Beckey and co-workers in 1969. It uses a localized electric field to form radical cations and anions, and sometimes secondary, protonated, or deprotonated, molecules such as $[M+H]^+$ and $[M-H]^-$. The process of ionization occurs at the tip of a needle electrode under vacuum. The needle comes into contact with the analyte by condensation of the vaporized sample after it has been heated in the source under vacuum.

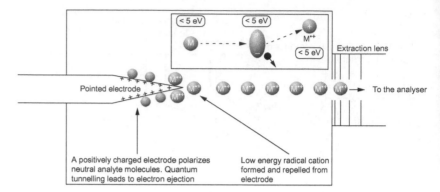

Figure 2.9 A general schema showing the process of field ionization (FI) formation of a radical cation.

When a strong electric field is applied (typically from a 5 kV supply), the region around a sample needle tip experiences a focused positive potential. Neutral analyte molecules that are in contact are polarized, and electrons are attracted to the positively charged needle surface. Quantum tunnelling is thought to occur, leading to the migration of an electron and its subsequent neutralization on the anode producing a positively charged radical cation. The newly formed cation experiences an immediate repulsion from the positively charged needle and moves out into the vacuum of the ion source, where it is then extracted towards the mass analyser by a counter-electrode. Figure 2.9 shows a schematic of this process.

FI usually produces little or no fragmentation. Positively or negatively charged radical ions when a positive or negative potential is applied to the needle, respectively. Subsequent ion–molecule interactions can occur when newly formed radical ions interact with a neutral molecule, leading to proton transfer and the formation of protonated analyte molecules.

FI has now been in part superseded by other ionization techniques such as MALDI, but it is still used, in particular for its ability to form molecular ions from compounds which do not form a molecular ion by EI or CI. Very little energy is transferred during ion formation, but the technique suffers from low sensitivity and the potential for thermal breakdown of analytes during their evaporation prior to ionization. FI has a sister technique called field desorption (FD) (Section 2.4), which utilizes a very similar ion source and ion formation mechanism. The needle, however, is precoated with the sample, and the process is therefore not one in which vaporized samples are analysed.

2.4 Desorption ionization techniques

Desorption ionization techniques represent a variety of ion sources that have transformed mass spectrometry. They are characterized by rapid addition of energy to a sample in the condensed phase, leading to collision-induced ionization and release of newly formed ions into the gas phase. They typically enable the analysis of non-volatile and thermally labile compounds via formation of even electron, protonated, or deprotonated molecules such as $[M+H]^+$ and $[M-H]^-$.

2.4.1 Matrix assisted laser desorption ionization (MALDI)

Matrix assisted laser desorption ionization (MALDI) is one of the best known and most widely used desorption ionization techniques. Developed in the 1980s it is characterized by a photon-induced ionization process that leads to energy and proton transfer via an intermediate crystalline matrix to yield desorbed ions. This is a soft-ionization process that tends to form singly charged positive ions, but deprotonated negative ions can be produced in negative mode. MALDI is highly sensitive and effective up to molecular weights well above 150 kilodaltons (kDa). It is therefore often used for the identification of proteins, protein modifications, peptides, and oligonucleotides.

2.4.1.1 MALDI: principles of ion formation

The mechanism of ionization starts with a pulsed transfer of laser energy to a co-crystallized mixture of analyte and a matrix. The matrix is typically a small organic aromatic acid which is strongly absorbing at the wavelength of the laser. This leads to rapid local heating, expansion, and vaporization of the mixture forming a small local gas plume. At this stage ionization of the matrix takes place. Typically, the matrix is in significant excess compared to the analyte (the ratio may be in the order of 1:1000). As a result, much of the laser energy is absorbed by matrix molecules, leading to formation of a range of electronic excited states and increased molecular collisions due to the higher local gas pressure in the newly formed plume. A detailed understanding of the MALDI ionization process has yet to be elucidated, as much of it takes place during the short-lived gas plume, but a number of models have been proposed. These include the photoionization and photochemical model, the polar fluid model, the solid state model, the dynamic model, the energy pooling model, and the proton transfer model. It is beyond the scope of this book to describe these models, but it is generally recognized that a number of the mechanisms that underpin these models are likely to occur in parallel.

Once primary matrix ions have formed in the gas plume, secondary ion-formation can occur. As the gas plume expands, local temperatures can be in the order of 600 K to 1200 K. Molecules and ions have high thermal energy, and primary ions react with neutral molecules that include the analyte, leading to proton transfer and formation of secondary ions observed in the mass spectrum. As the gas plume continues to expand, the pressure reduces rapidly and ion–molecule collisions become less frequent, enabling newly formed analyte ions to be extracted into the mass spectrometer.

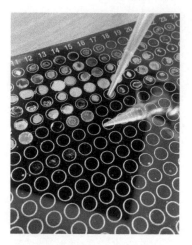

Figure 2.10 A matrix assisted laser desorption ionization (MALDI) plate being 'spotted' with a matrix/analyte mixture.

2.4.1.2 **MALDI: source design and function**

The MALDI ionization source combines sample introduction with the ionization process. The sample is usually mixed with a matrix in solution and co-crystallized on a specific 'spot' on a MALDI plate. Figure 2.10 shows a MALDI plate with analyte/matrix spots prepared for analysis, and Table 2.3 lists common matrices and their suitable solvents. Once the spots are dry, the plate is inserted into the source region of the mass spectrometer via a vacuum interlock.

The choice of matrix, how efficiently it is mixed with the sample, and how well it has been dried are important factors for optimal ionization performance. Once inside the mass spectrometer the MALDI plate is aligned with a laser (which is commonly a solid-state, CO_2 (IR), or nitrogen (UV) laser). Variables that can be used to optimize the spectrum include laser energy, where on the spot the laser is fired, and the number of laser pulses fired on a particular spot. Ions generated are extracted from the area on the plate via a potential applied between the target plate and an extraction lens. Matrix ions are in significant excess and are therefore inevitably extracted along with sample ions, leading to a mass spectrum dominated in the low-mass region by matrix ions. Instrument control software

Table 2.3 Table of common matrices used in matrix assisted laser desorption ionization (MALDI), including suitable solvents and analytes.

Matrix	Abbr.	Proton affinity (kJ mol^{-1})	Suitable solvents	Applications
Sinapinic acid	SA	860	MeCN, H$_2$O, acetone, chloroform	Proteins Peptides Lipids Polar polymers
Alpha-cyano-4-4-hydroxycinnamic acid	CHCA	841	MeCN, H$_2$O, ethanol, acetone	Peptides
2,5-Dihydroxybenzoic acid	DHB	703	MeCN, H$_2$O, MeOH, acetone, chloroform	Nucleotides Oligonucleotides Oligosaccharides
Glycerol	–	961	–	Proteins, peptides (liquid matrix)
Dithranol	–	697	Dichloromethane, acetone, ethanol	Lipids Non-polar polymers
Nicotinic acid	NA	829	MeOH, H$_2$O, MeCN	Peptides Proteins
3-Hydroxypicolinic acid	3-HPA	766	Ethanol	Oligonucleotides Oligosaccharides
2,4,6-Trihydroxyacetophenone	THAP	891	MeCN, H$_2$O, ethanol	Peptides, Nucleotides Carbohydrates Oligonucleotides

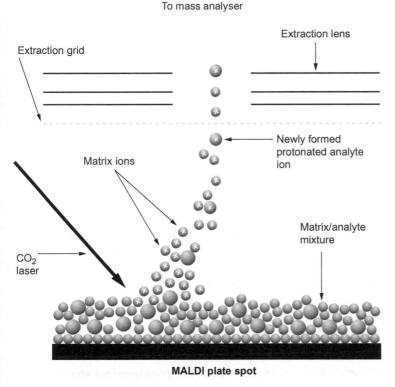

To mass analyser

Extraction lens

Extraction grid

Newly formed
protonated analyte
ion

Matrix ions

Matrix/analyte
mixture

CO_2
laser

MALDI plate spot

Figure 2.11 A schematic showing the layout of a matrix assisted laser desorption ionization (MALDI) ion source and the formation of protonated analyte molecules via matrix ionization.

often automatically removes the region of the mass spectrum below 300–400 Da to eliminate 'matrix noise'. Figure 2.11 shows a schematic of the MALDI ion source and protonation of analyte molecules by matrix ionization.

2.4.1.3 MALDI: performance and applications

Because proton transfer is the primary mechanism of ionization in MALDI, molecular weight is not a significant bias, and relatively high molecular weight compounds can be protonated or deprotonated. The matrix ions found at low mass can induce significant signal suppression for low molecular weight compounds, and MALDI is therefore not usually used for analysis below approximately 500 Da. The dispersal of analyte molecules and efficient energy transfer between matrix and analyte leads to relatively high ionization efficiency and analytical sensitivity which can be down at the pico-molar level. Compared to electrospray ionization (see Section 2.5.1.4), MALDI is more tolerant of salts, buffers, and ionic contaminants, but these components can still affect sensitivity. Quantitative analysis using MALDI can be particularly challenging and is not commonly performed. Careful choice of internal standards is essential if this is to be attempted (see Chapter 7).

Relatively recently the development of 'MALDI imaging' has taken advantage of the precision and specificity of the laser-initiated ionization process. In this application, mapping of molecular features across the surface of a sample (often a biological tissue) is achieved by the mobility and high resolution of the ionizing laser. This can provide spectra representing the distribution of compounds on the surface of a sample. Applications include, for example, mapping the differential expression of proteins across tissues and the uptake of chemotherapies across tumours (see also Section 8.5.1).

The main disadvantages of MALDI are that it is much less amenable to the analysis of small molecules compared to other ionization techniques and it is generally challenging to produce quantitative results. Spectral peak resolution is also lower than for soft ionization techniques such as electrospray ionization (discussed in Section 2.5). The advantages of MALDI include its ability to ionize very high molecular weight compounds and the formation of singly charged ions even at high mass. It is robust to matrix additives and, in particular, it is easy to use and sample analysis is fast. These characteristics have made MALDI a transformative tool in a number of research areas including protein science and polymer chemistry. It remains a hugely important and commonly applied ionization source in mass spectrometry.

2.4.2 Secondary ion mass spectrometry (SIMS) and fast atom bombardment (FAB)

2.4.2.1 SIMS

The first SIMS techniques were developed in the 1940s and 1950s, enabling highly sensitive surface analysis based on a desorption-ionization mechanism. Today they are generally used for the analysis of solids by mass spectrometry, and applications include analysis of semiconductors, glass, stainless steel, meteorites, fuel cells, museum artefacts, alloys, and biomaterials. The latter has become increasingly popular, and there are now applications in many biological fields, including medicine, cell, and tissue studies.

Secondary ions are produced from a high-energy primary ion beam (often caesium ions accelerated to 20–30 kV), which bombard the surface of a sample material. Secondary ions are formed via charge transfer as well as displacement of ions from the surface of the material by momentum transfer. Once ejected from the surface, ions are extracted towards the mass analyser by an electrical potential. SIMS is generally not as sensitive as MALDI or electrospray ionization, and has a more limited mass range in the order of 300–13,000 Da. The ability to focus the primary ion beam enables extremely high-resolution sampling from solid surfaces, and the formation of a 'chemical map' down to the low micron level is possible. High spatial resolution and fast analysis times make SIMS particularly useful for materials science and surface chemistry applications. Figure 2.12 provides a schematic of the ionization process that takes place in a SIMS ion source.

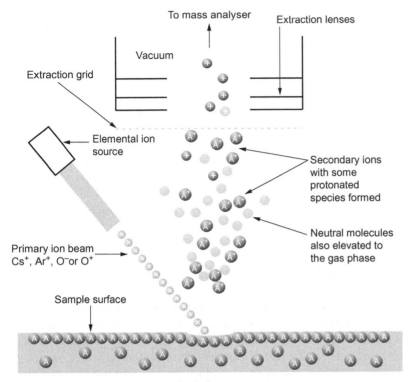

Figure 2.12 A schematic showing the generic layout of a secondary ion mass spectrometry (SIMS) ion source and the formation of secondary ions.

2.4.2.2 FAB

FAB ionization was developed in the 1970s as a variant of the SIMS approach, and led to fast and efficient analysis of proteins and peptides for the first time. It is now largely superseded by electrospray ionization and MALDI, due to their higher sensitivity and extended mass range. FAB, like SIMS, is based on a secondary ion formation process. Two major differences compared to SIMS are that FAB uses a neutral atom beam to create secondary ions (often xenon or neon atoms) and a 'liquid matrix', rather than solid surface, is analysed. The liquid matrix is usually a low-volatility solvent in which the sample is dissolved, and ionization takes place at the surface of the liquid where the beam interacts with the solute. As the analyte molecules are in solution and in motion, once ionized and extracted by an electric field they are quickly replaced, and hence relatively long acquisition times can be used with FAB. The mass range and sensitivity of FAB do not approach that of ESI or MALDI, however. Despite being very little used today, at the time of its development FAB transformed the ability of mass spectrometry to analyse biological molecules and has made a major contribution to the field of protein mass spectrometry.

2.4.3 Field desorption (FD)

FD ionization is related to FI (described in Section 2.2) and uses virtually the same ion source design and mechanism of ionization. The way in which the sample is introduced into the ion source is, however, different. The sample is coated on to a needle electrode and allowed to dry. A localized electric field is then applied to the needle under vacuum, which leads to desorption of the analyte as part of the ionization process, forming radical cations and anions and sometimes secondary, protonated, or deprotonated molecules, using a very similar mechanism to FI. FD produces even less fragmentation, however. Subsequent ion–molecule interactions can also occur in which the newly formed radical ion interacts with a neutral molecule, leading to proton transfer and the formation of protonated molecules. Gentler still than FI, FD is one of the 'softest' ionization techniques available that utilizes an electron ionization mechanism.

2.5 Atmospheric pressure ionization techniques

The importance of ESI (and MALDI) was acknowledged by the 2002 Nobel Prize for Chemistry shared by John B. Fenn and Koichi Tanaka, who independently and respectively pioneered the development of these techniques for mass spectrometry. 'To give molecular elephants wings' is a fragment of John Fenn's acceptance speech referring to the ability of ESI to enable even extremely large proteins to be ionized and measured by mass spectrometry with relative ease.

The introduction of atmospheric pressure ionization (API) techniques in the 1980s had a profound effect on mass spectrometry, enabling new and wide-ranging applications. This group of ionization methods generally forms protonated or deprotonated molecules, or adducts with other anions or cations in a source at atmospheric pressure. These techniques can be differentiated by the mechanism through which gas-phase ions are produced. Electrospray ionization (ESI) is the most commonly used API technique, but others have specific applications for which they are particularly well suited.

2.5.1 Electrospray ionization (ESI)

The principles of ESI are based on the rapid evaporation of solvent droplets containing dissolved analyte molecules, in a strong electric field. ESI can be used to ionize a very wide range of organic and biological compounds, and is easily interfaced with liquid chromatography (LC) systems at flow rates ranging from nano-litres/minute to > 1 mL/minute (see Chapter 7). Due to its flexibility and broad analyte coverage, ESI is one of the most commonly used ion sources in modern mass spectrometry, particularly for the analysis of complex environmental and biological molecules.

2.5.1.2 ESI: principles of ion formation

Described by Sir Geoffrey Ingram Taylor in 1964, the 'Taylor cone' refers to the shape adopted by an electrically conducting liquid, such as an aqueous droplet containing electrolytes, when it is exposed to an electric field. The shape of the droplet deforms from that determined by surface tension alone and produces a conical shape as the electric field increases. A Taylor cone is commonly observed at the tip of the capillary during electrospray ionization.

Analytes, dissolved in a suitable solvent, are pumped (either from an LC system or directly from a sample reservoir) through a stainless steel capillary held at high electrical potential difference. A droplet emerging from the capillary, in positive ion mode, experiences the electric field. This polarizes solvent molecules, aligning dipolar water molecules and initiating the migration of positively charged electrolyte ions towards the meniscus surface of the capillary tip, forming a specific shape due to the electric field known as a 'Taylor cone'. Figure 2.13 provides a schematic of Taylor cone formation at the end of a capillary during electrospray ionization.

Elongation of the tip of the cone leads to the formation of droplets with excess positive charge. Once they emerge they are attracted towards the mass spectrometer inlet. This attraction is enhanced by the reducing positive electric field, but the droplets with high positive charge also repel each other. These dynamics manifest in the formation of a small jet at the capillary tip which fans out rapidly into many small independent droplets. The tiny droplets evaporate quickly, often promoted by the introduction of a drying gas and heating. Surface tension keeps the droplets intact initially, but as the droplets become smaller, positively charged electrolyte ions in particular, experience an increasing coulombic repulsion. When this repulsion reaches the *Rayleigh limit*, overcoming surface tension, the micro-droplets 'explode' into many smaller nano-droplets. This process of evaporation and fission continues until analyte ions are ejected from the droplet (the so-called ion evaporation model, IEM), or a single analyte ion enters the gas phase after complete evaporation of the remaining droplet solvent (the so-called charge residue model, CRM). During the ESI process, adduct formation occurs, usually with one or more protons associated at one or more basic sites of the molecule. This process produces positively charged protonated (or cationized) species. Reversing the polarity of the ion source promotes proton abstraction from an acidic site (or sites) to form negatively charged, deprotonated species. Table 2.4 provides a list of common adducts that can be formed during electrospray ionization.

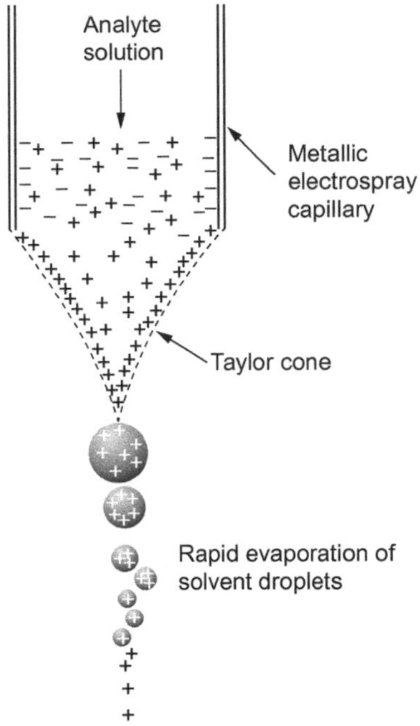

Figure 2.13 Taylor cone formation during electrospray ionization.

2.5.1.3 ESI: source design and function

The stainless steel capillary is 40–60 µm in diameter and has a voltage applied, usually 1–4 kV depending on the ion-mode and flow rates used. The exit of the capillary is typically located a few centimetres from the mass spectrometer inlet.

Table 2.4 A list of common adducts encountered in electrospray ionization. Often a mixture of adducts can be formed in complex samples for a single analyte, but protonated or deprotonated molecules are usually predominant in positive and negative ion modes respectively.

Common positive ion adducts	Ion	Common negative ion adducts	Ion
$[M+H]^+$	Proton	$[M-H]^-$	Proton (removal)
$[M+Na]^+$	Sodium	$[M+Cl]^-$	Chloride
$[M+K]^+$	Potassium	$[M+HCO_2]^-$	Formate
$[M+H+CH_3CN]^+$	Acetonitrile	$[M-H+C_2H_6OS]^-$	DMSO
$[M+NH_4]^+$	Ammonium	$[M+C_2H_3O_2]^-$	Acetate

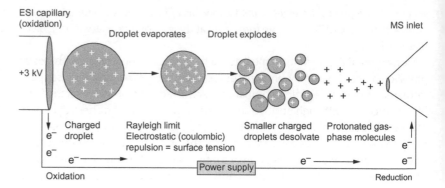

Figure 2.14 A schematic showing the layout of an electrospray ionization source in positive ion mode and the formation of protonated analyte molecules through droplet evaporation.

A coaxial sheath gas (N_2) flows around the outside of the capillary to assist nebulization of the sample solution exiting the capillary. Although ESI takes place at atmospheric pressure, outside the mass spectrometer, the ESI source often has a sealed housing to aid localization of heating and introduction of a counter current of N_2 gas to help the evaporation and desolvation process, which is particularly important at higher flow rates. The housing also prevents release of solvent vapour into the laboratory. In general, the ionization efficiency of ESI is several orders of magnitude greater than an electron ionization source. Figure 2.14 shows a schematic of a general ESI ion source and the process of ion formation that takes place.

2.5.1.4 ESI: performance and applications

The ease with which ESI can be coupled directly to liquid chromatography provides many advantages over other ion sources for LC-MS analysis (Chapter 7). This makes ESI particularly suitable for a wide range of applications, including environmental analysis, toxicology, drugs of abuse and doping control, pesticide analysis, and biological applications, including the analysis of metabolites, proteins, peptides, and oligonucleotides (see Chapter 8). Indeed, the development of the field of proteomics by mass spectrometry in the last twenty-five years has been heavily reliant upon of electrospray ionization. Unlike MALDI, ESI readily forms multiply charged ions, sufficiently low in energy to enable ionization of proteins and other large oligomers without fragmentation. Figure 2.15 shows the mass spectrum of the protein myoglobin produced by electrospray ionization, showing a characteristic 'charge state envelope'.

The signal response from ESI is analyte and sample composition dependent. It is determined by the ability of functional groups present in molecules to accept or donate a proton under electrospray conditions, but is also susceptible to interferences from other molecules in the sample matrix. 'Matrix effects' are interferences that additional compounds in a sample have on the ion signal of the analyte. They are often associated with the analysis of biological samples, although exogenous contaminants (such as plasticizers from samples coming

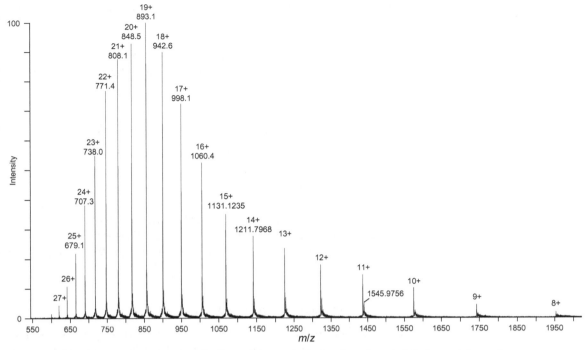

Figure 2.15 The electrospray mass spectrum for the protein myoglobin showing a characteristic charge state envelope.

in contact with plastic-ware) can lead to matrix effects in the analysis of otherwise 'purified' compounds. Matrix effects generally reduce (but can occasionally enhance) analyte ionization efficiency. A common type of matrix effect, which leads to attenuation of signal response, is known as 'ion suppression'. This can have a significant negative effect on sensitivity and the ability to quantify analyte concentrations which can make abundance comparisons between analytes challenging (see Chapter 7 for further discussion of matrix effects and ion suppression). This makes sensitivity in ESI very dependent on chemical structure and on mobile phase composition, salt content, and composition of additives. Somewhat counterintuitively, sensitivity generally increases with a reduction in flow rate through the capillary. The efficiency of ion formation is low (< 0.1%) for standard flow ESI (0.1–1 mL/min), but is, highly sensitive relative to other ionization techniques and can provide analysis of certain compounds down to attomolar sample concentrations (10^{-18} moles/L).

2.5.1.5 Nano-ESI

Nano-electrospray (nano-ESI) is a low-volume and low-flow version of standard ESI which capitalizes on the enhanced efficiency and sensitivity that can be achieved at nanolitre/min flow rates. The efficiency of nano-ESI is dramatically higher and can be higher than 40% in some cases when compared to standard

ESI, making it one of the most efficient ionization processes available for mass spectrometry. Nano-ESI comes with some practical challenges, however, such as maintaining stable flows and being prone to blockages. It is used commonly in proteomics in conjunction with nano-LC-MS where the low sample volume requirements coupled with high sensitivity make it particularly useful for the analysis of proteolytic digests.

2.5.2 Atmospheric pressure chemical ionization (APCI)

APCI was developed in the 1970s, and exhibits some similarities to CI. It predominantly utilizes a proton transfer, 'soft' ionization process that is initiated from the ionization of a reagent gas. Unlike conventional CI, it takes place at atmospheric pressure and can be easily coupled to liquid chromatography for the analysis of complex samples. It is efficient for the analysis of non-polar molecules where ESI is less effective. APCI is much less efficient for the analysis of compounds at higher masses compared to ESI, and has an effective upper limit of around 1500 Da.

2.5.2.1 APCI: principles of ion formation

In the process of ion formation a heated nebulizer converts a liquid eluent stream into gas-phase molecules by rapid heating. As this takes place at atmospheric pressure there is an excess of N_2 and some O_2, which become 'reagent gases', leading to a process reminiscent of CI using ammonia or methane. Reagent gas molecules are in significant excess compared to the analyte, and are primarily ionized by electron ionization initiated by a sparking coronal discharge located next to the source inlet. In positive ion mode radical cations are produced which then rapidly interact with neutral solvent molecules in the gas phase, forming intermediate protonated species. These subsequently undergo proton transfer reactions with the desolvated molecules, and protonated analyte ions are extracted into the mass spectrometer via a conventional sampling cone. Figure 2.16 represents the ion formation processes that can occur during atmospheric pressure chemical ionization.

2.5.2.2 APCI: source design and function

APCI source design is similar to that of ESI, but a neutral rather than charged spray is produced from the probe. An analyte solution is introduced via a syringe pump or liquid chromatography inlet. A droplet spray is generated in the source and nebulized via a nitrogen sheath gas. The whole system is heated at the point of nebulization to about 120 °C, which leads to the formation of mostly gas-phase solvent and analyte molecules. The controlled temperature and gas flow ensure that these solvent molecules evaporate quickly as in ESI, but no electric field is applied so they do not contain excess charge. A coronal discharge needle is located close to the sampling cone of the mass spectrometer, which is used to produce

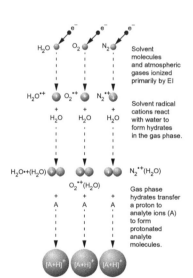

Figure 2.16 Simultaneous ion formation processes that can occur during atmospheric pressure chemical ionization (APCI).

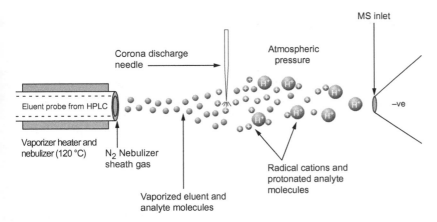

Figure 2.17 A schematic showing the layout of the atmospheric pressure chemical ionization (APCI) ion source and the formation of protonated analyte molecules.

high-energy electrons for radical cation formation. Figure 2.17 provides a schematic of the ionization process taking place during APCI.

2.5.2.3 APCI: performance and applications

APCI is commonly interfaced with high-performance or ultra-high-performance liquid chromatography (HPLC or UPLC, see Chapter 7). Unlike ESI, APCI ionization efficiency is much less affected by eluent flow rates, and matrix effects tend to be less severe. APCI can therefore often be used more effectively at higher flow rates than ESI, which makes it more efficient for coupling to conventional HPLC. Applicable flow rates are in the range 200 μL–2000 μL/min. Although efficiency is less affected by flow rate, there is more thermal degradation and fragmentation than for ESI. The mass range is also significantly reduced, making it more suitable for the analysis of low polarity, low molecular weight compounds. Targeted pharmaceutical and metabolite analyses, including the analysis of lipids, are common APCI applications.

2.5.3 Atmospheric pressure photoionization (APPI)

APPI uses high-energy photon irradiation to ionize molecules in the gas phase. It is similar to APCI in that it uses a nebulized spray to create gas-phase analyte molecules but no electric discharge or electric field is used to promote ion formation. Instead, UV light transfers energy to the analyte. As for many API sources it can be coupled to liquid chromatography, but there is a narrower range of molecules for which APPI provides adequate sensitivity. It can, however, be sensitive to selected compounds, particularly those with low polarity. Because of this selectivity, APPI is generally less susceptible to ion suppression and contamination than are other API sources.

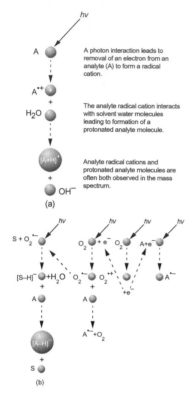

Figure 2.18 (a) Formation of positive ions by direct atmospheric pressure photoionization (APPI). (b) Formation of negative ions by direct APPI.

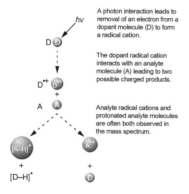

Figure 2.19 Formation of positive ions by dopant-assisted atmospheric pressure photoionization (APPI).

2.5.3.1 APPI: principles of ion formation

The process starts in a similar way to APCI with nebulization and vaporization of a solvent stream containing the analyte. Irradiation of the vapour with UV radiation results in absorption of high-energy photons. If the energy of the photon is higher than the ionization energy of the molecule, an electron will be ejected, leading to radical cation formation. However, this mechanism probably accounts for only a small percentage of the ions formed, for two reasons. First, the number of analyte molecules will almost always be low compared to solvent and gas molecules, leading to little direct interaction between photons and the analyte at atmospheric pressure. Second, protonated molecules are often found in excess to radical cations, suggesting that an alternative ionization mechanism also takes place. It is thought that a number of additional mechanisms can occur that include charge exchange with neutral analyte molecules as well as proton transfer reactions.

APPI can form negative ions under appropriate conditions, and three separate mechanisms have been identified: simple electron capture by thermal electrons produced from the photoionization of a dopant, charge exchange from the reaction of analyte molecules with photoionized dioxygen, and proton transfer from basic solvent species or other gas-phase ion molecule interactions (Figures 2.18 and 2.19). Dopants are compounds such as toluene, acetone, or chlorobenzene that enhance analyte ionization. They are introduced into the ionization chamber where UV radiation readily ionizes these molecules, leading to the formation of free radicals and molecular ions which subsequently ionize analyte molecules by electron and proton transfer reactions. The mechanisms of ionization that take place can be varied and depend upon the type of solvent, dopant, and, to a significant extent, the concentration of the analyte ions. Although basic mechanisms can be predicted, a detailed understanding of the processes taking place during APPI remains to be elucidated.

2.5.3.2 APPI: source design and function

The APPI source is very similar to APCI, but the coronal discharge needle is replaced by a photon source, most commonly a hydrogen or krypton discharge UV lamp. The latter is usually capable of emitting photons at 10.0 and 10.6 eV. A liquid chromatography eluent stream is interfaced with a sprayer which is comprised of a heater and a sheath gas that nebulizes and heats the eluent to quickly vaporize it in the ion source chamber. As for ESI, inline or orthogonal sprayer orientations can be used, and the temperature and gas flow are controlled to ensure that solvent and analyte molecules are efficiently vaporized. The mass spectrometer inlet is positioned close to the gas cloud, and newly formed ions are brought into the high vacuum of the mass spectrometer via diffusion and electrostatic extraction (Figure 2.20)

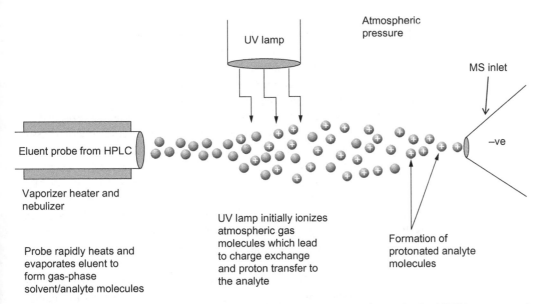

Figure 2.20 A schematic showing the general layout of an atmospheric pressure photoionization (APPI) ion source and the formation of protonated analyte molecules.

2.5.3.3 APPI: performance and applications

Although ionization of low-polarity and non-polar compounds is where APPI provides unique capabilities, the majority of applications have tended to be focused on molecules for which ionization potential overlaps with APCI and ESI (see Section 2.5.1 and 2.5.2). These include pharmaceutical, clinical, and environmental applications. For example, APPI has been particularly effective for the analysis of steroids, sterols, and other hydrophobic compounds in complex matrices such as blood plasma. The selectivity and potential sensitivity of APPI for specific compounds has led to environmental monitoring such as the analysis of oestrogens, perfluorooctane sulfonates (PFOS), and phenylbenzotriazole mutagens (PBTAs) in contaminated water. APPI provides superior sensitivity for some air quality monitoring applications, including the analysis of polycyclic aromatic hydrocarbons (PAHs) and ketones from automobile combustion emissions. APPI remains, however, a rather specialized ionization technique.

2.6 Ambient ionization techniques

For almost all the ionization techniques described so far, we have seen that specific sample preparation approaches are required in order to introduce the sample into the mass spectrometer. For example, ESI requires analytes to be dissolved in an appropriate solvent. MALDI requires co-crystallization of the sample with a matrix on a plate that must be introduced into the mass spectrometer. In some cases these particular sample preparation requirements can be an impediment to the analysis of certain types of sample, especially 'real world' samples where analysis *in situ* is preferable or indeed essential.

Since the early 2000s the development of so called 'ambient ionization techniques' has been rapid. These typically function under atmospheric pressure without the need for transformative sample preparation. Most of them incorporate conventional ionization mechanisms such as those used in API or laser desorption ionization techniques, but they enable analysis of a very wide range of analytes present, for example, on inanimate or biological objects such as banknotes, mould, skin, or the surface of fruits, cells, and tissues. There are now well over 100 differently named approaches, and a selection of these can be found listed, along with the types of sample for which they are amenable, in Table 2.5.

Although many ambient ionization techniques take advantage of unique physical processes, they are fundamentally based on common mechanisms generally found to take place in ESI, APCI, or laser desorption ionization. They

Table 2.5 A list of common ambient ionization techniques and suitable analytes.

Ambient ionization technique	Acronym	Types of analyte
Desorption electrospray ionization	DESI	Polar solids and liquids
Direct analysis in real time	DART	Polar and non-polar solids, liquids, and gases
Atmospheric pressure solids ionization probe	ASAP	Polar and non-polar solids and liquids
Secondary electrospray ionization	SESI	Polar liquids
Rapid evaporative ionization mass spectrometry	REIMS	Tissue
Desorption ionization by charge exchange	DICE	Polar and non-polar solids
Paper spray	PS-MS	Low- to medium-polar solids and materials
Laser ablation electrospray ionization	LAESI	Polar solids and liquids
Easy ambient sonic spray ionization	EASI	Polar solids and liquids
Desorption atmospheric pressure chemical ionization	DAPCI	Polar solids and liquids
Low-temperature plasma probe	LTP	Polar and non-polar solids, liquids, and gases
Atmospheric pressure penning ionization	APPeI	Polar and non-polar liquids and gases
Desorption atmospheric pressure photoionization	DAPPI	Polar and non-polar liquids
Laser spray ionization	LSI	Polar liquids
Electrospray laser desorption ionization	ELDI	Polar solids and liquids
Flowing atmospheric-pressure afterglow	FAPA	Polar and nonpolar gases
Radio-frequency acoustic desorption and ionization	RADIO	Peptides and proteins

generally exploit 'soft' ionization mechanisms, and predominantly produce radical cations or cation or anion adducts such as $[M+H]^+$ and $[M-H]^-$. They are characterized by:

- Direct analysis of the sample with little or no sample preparation.
- Ionization occurring at atmospheric pressure.
- Desorption of ions from the surface of samples.
- Hybrid or multiple mechanisms of ion formation.

One of the earliest ambient ionization techniques to be developed was desorption electrospray ionization (DESI) by Cooks, Takats, and co-workers in 2004. This process resembles electrospray ionization, but rather than the analyte being dissolved in the solvent spray, the charged droplets of solvent are directed at the surface of the sample. They strike sample molecules, momentum and charge are transferred, and newly formed ions are ejected from the sample surface and subsequently extracted towards the mass spectrometer inlet (Figure 2.21).

Another ambient ionization technique known as direct analysis in real time (DART) was developed around the same time as DESI, and is in some ways complementary. It is capable of ionizing non-polar, neutral molecules released into the atmosphere from a solid material by thermal desorption. Ions are formed by charge transfer from water vapour and other gas-phase ions at atmospheric

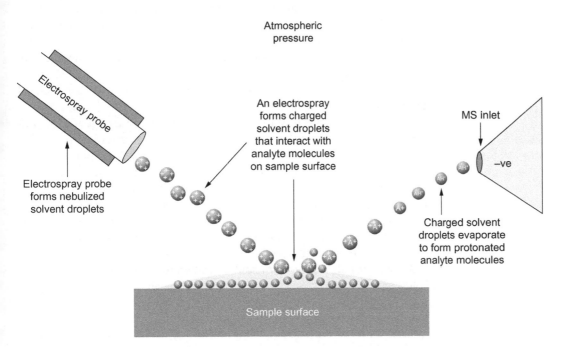

Figure 2.21 A general schematic showing the process of desorption electrospray ionization (DESI) ambient ionization to form protonated analyte molecules.

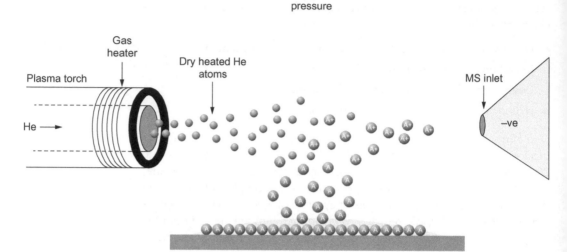

Figure 2.22 A schematic showing the process of direct analysis in real time (DART) ambient ionization to form protonated analyte molecules.

pressure. The process is initiated from a 'dry' stream of electronically excited neutral gas atoms (usually He) which are heated and directed at the sample (Figure 2.22). These atoms interact with atmospheric molecules and other atoms along the way, to form a range of charged species which subsequently transfer their charge to the analyte. Specific mechanisms are still not well understood, but appear to be highly dependent on the chemical properties of the analyte. Protonation and deprotonation, charge transfer, and Penning ionization are all thought to occur under suitable conditions.

DART is also particularly well suited to the analysis of compounds which have been deposited on the surface of materials. Forensic applications have included the detection of explosives, warfare agents, and volatile fingerprint residues. Environmental applications include cosmetics analysis, soils analysis, and now an increasing number of biological applications that include metabolomics and molecular imaging. Food quality and safety has seen many applications, including pesticide residue detection on fruits and vegetables. DART has also been used for clinical applications, including analysis of plasma and urine to identify specific inborn errors of metabolism.

A third example of a new ambient ionization technique is the atmospheric solids analysis probe (ASAP), which is used for the analysis of solids or liquids transferred or evaporated onto an inert surface. The ASAP rapidly heats (230–500 °C) vaporizing samples using a hot nitrogen gas flow, and ionization is induced by a coronal discharge in a similar way to APCI (Figure 2.23). Rapid

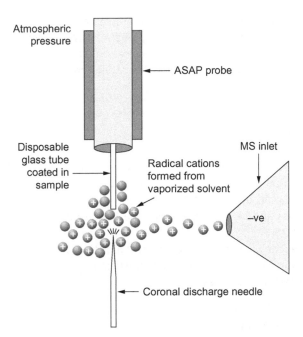

Figure 2.23 A schematic showing the process of atmospheric solids analysis probe (ASAP) ambient ionization forming radical cations.

analysis of samples is possible using a disposable glass tip which is usually dipped into liquid or solid samples just prior to analysis. A range of ionization mechanisms are thought to occur depending on the sample type. For example, if water vapour is present (if the sample is not sufficiently dry), protonation or deprotonation can occur, forming $[M+H]^+$ or $[M-H]^-$ species. If the sample and source are 'dry', then radical cations are formed predominantly via coronal discharge. Forensic applications have included analysis of bank notes and inks for forgery analysis, but reaction mixtures in the organic chemistry laboratory can also be easily sampled to monitor reaction products using ASAP.

The mechanisms of ionization for most of the ambient ionization techniques are challenging to investigate, and ion/charged molecule formation processes are generally less well understood than for some of the ionization techniques discussed. Limits of detection can be low in the parts per billion (ppb) or femto-mole range (10^{-15} moles), but ion intensities can often be highly dependent on the sample surface, its temperature, moisture content, and other factors which can be difficult to control. Quantitative experiments using ambient ionization techniques have seldom been pursued, and matrix effects have been shown to have a substantial impact on the attenuation of ion signals in a number of studies. However, the advantages associated with ambient ionization for *in situ* analysis, particularly for environmental, biological, pharmaceutical, and forensic applications, have ensured their place in the expanding repertoire of ion sources available to mass spectrometry. Minimal sample handling reduces sample loss and distortion of natural chemical fingerprints, making ambient techniques particularly appropriate for portable mass spectrometers. These have now been developed for applications sampling the natural environment on earth and in space.

2.7 Summary

From the material presented in this chapter you should now be familiar with the main ionization techniques and sources used in modern mass spectrometers. You should understand the different physical mechanisms by which each operates and be able to describe the basic processes of ion formation that take place.

From the following list of ionization techniques you should also be able to sketch the ion source design and suggest which may be most suitable for the analysis of particular analytes.

- Electron ionization (EI)
- Chemical ionization (CI)
- Electrospray ionization (ESI)
- Secondary ion mass spectrometry (SIMS)
- Fast atom bombardment (FAB)
- Inductively coupled plasma ionization (ICP)
- Atmospheric pressure chemical ionization (APCI)
- Field ionization (FI)
- Field desorption ionization (FD)
- Matrix assisted laser desorption ionization (MALDI)
- Atmospheric pressure photoionization (APPI)
- Desorption electrospray ionization (DESI)
- Direct analysis in real time (DART)
- Atmospheric solids analysis probe (ASAP)

You should be aware of some of the main advantages and disadvantages of each ion source, including the mass range within which each operates and their relative sensitivities to different types of analyte.

2.8 Exercises

2.1   Draw a generic equation for the formation of a charged species from a neutral molecule by electrospray ionization in (i) positive ion mode and (ii) negative ion mode.

2.2   For each of the following analytes, name an ionization technique which may be suitable for its analysis by mass spectrometry.

- Propane
- Methyl stearate
- L-phenylalanine
- Magnesium
- Bovine serum albumin (a protein of molecular mass 67 kDa)

2.3   Name two ambient ionization techniques and state two similarities and two differences in the process of ion formation.

2.4   List three similarities and three differences between electrospray ionization and atmospheric pressure chemical ionization.

2.5   Draw a generic equation for the formation of a charged species from a neutral molecule by matrix assisted laser desorption ionization.

2.6   Which ionization sources are most suitable for the analysis of non-polar compounds?

2.7   Provide a general equation for the process of radical cation formation using an electron ionization source, and state the typical energy that electrons are accelerated to in a standard electron ionization source.

2.8   Explain the differences between 'hard' and 'soft' ionization.

2.9   Which atoms within a neutral molecule are most likely to be protonated in positive electrospray ionization?

2.10 What property of a sample can lead to quite different ion abundances for analysis of the same analyte at the same concentration using electrospray ionization?

2.9 Further reading

Chapman, J. R. (1993). *Practical Organic Mass Spectrometry*, 2nd edn. Chichester: Wiley.

Cole, R. B. (ed.) (2010). *Electrospray and MALDI Mass Spectrometry*, 2nd edn. New York: Wiley.

Hoffman, E. de and Stroobant, V. (2007). *Mass Spectrometry: Principles and Applications*, 3rd edn. Chichester: Wiley.

Munson, M. S. B. and Field, F. H. (1966). 'Chemical ionization mass spectrometry I. General introduction', *J. Am. Chem. Soc.* 88 (12), 2621–30.

Takats, Z., Wiseman, J. M., Gologan, B., and Cooks, R. G. (2004). 'Mass spectrometry sampling under ambient conditions with desorption electrospray ionization', *Science*, 306, 471–3.

Watson, J. T. and Sparkman, O. D. (2007). *Introduction to Mass Spectrometry*, 4th edn. Chichester: Wiley.

3 Methods of mass analysis

3.1 Introduction

The mass analyser is at the heart of every mass spectrometer. It is the device that separates ions according to their m/z ratios. Mass analysers can be divided into two main classes: beam analysers and trapping analysers. All analysers utilize electric or magnetic fields to manipulate ions in a way that allows their m/z to be measured. Sometimes this can be as simple as acceleration by a potential difference, followed by measurement of time of flight over a fixed distance, or as complex as the motion of ions in a quadrupolar electric field. Whatever the method, a reliable correlation between the measured behaviour of the ion and its m/z value is essential. With appropriate calibration, modern commercial mass spectrometers output the mass spectrum automatically on the m/z scale.

In this chapter we will examine the basic principles of operation of the most common analyser types. The physical and mathematical treatment is necessarily simplified in some areas, whilst maintaining sufficient detail to allow a reasonable understanding of how they function. All the analysers described here require a high vacuum for efficient operation. The precise vacuum requirement is analyser-dependent, and can span the range 10^{-3} to 10^{-10} mbar.

Different analyser types possess different properties and performance characteristics, including size, sensitivity, resolving power, m/z range, scan time, duty cycle, and cost. These are all important considerations in selecting the most appropriate mass spectrometer for a particular analytical task.

Table 3.1 Types of mass analyser.

Beam analysers	Trapping analysers
Magnetic sector	Ion trap
Quadrupole	Ion cyclotron Resonance
Time-of-flight	Orbitrap

3.2 Magnetic sector analysers

The magnetic sector is probably the mass analyser design most familiar to chemistry students. This is largely for historical reasons as, in fact, the use of sector instruments has declined significantly in recent years. Nevertheless, it is still important and instructive to examine the principles upon which these analysers function. Figure 3.1 shows the schematic layout of a simple magnetic sector MS. Ions generated in the source are accelerated into the region between the poles of a magnet by a potential, V. The magnetic field B

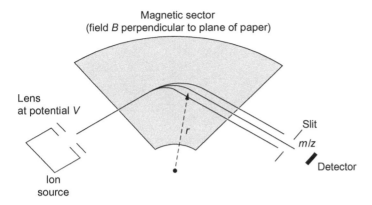

Figure 3.1 Layout of a mass spectrometer using a magnetic sector mass analyser.

(perpendicular to the plane of the paper) causes ions of m/z to deflect with a radius r through a slit and onto a detector. Note that only ions with the correct m/z value are detected. Ions of different m/z can be focused onto the detector by adjusting either B, V, or r. The most common method is to change B by using an electromagnet and scanning the applied current. This is termed a *magnet scan*. *Voltage scanning* at constant B is also possible, but is less desirable over a wide m/z range, since, at low values of V, sensitivity may be adversely affected. This is due to the relatively low kinetic energy, and resulting low transmission efficiency, of the ions when V is reduced. Different values of r may be sampled by keeping B and V constant and using an array detector, or a number of Faraday cups at discrete values of r. This latter strategy is commonly employed in isotope ratio mass spectrometers, as it allows the simultaneous, and therefore very precise, measurement of isotopic abundance (for example, $^{13}C/^{12}C$).

A magnet scan over the range m/z 50–1000 is typically achieved at rates of the order of 1 scan s^{-1} and a duty cycle of 0.1%. If the magnet is set to transmit ions of a single m/z value, then a duty cycle of 100% is achieved. The practical upper m/z limit for a magnetic sector MS depends upon the size and geometry of the magnet, but is approximately 2000–5000.

The relatively long path lengths in sector instruments require the combination of a high accelerating potential V (typically 5–10 kV) and a high vacuum (10^{-7} mbar) to achieve good ion transmission and associated sensitivity.

3.2.1 Double focusing sector analysers

The resolving power of a magnetic sector mass analyser can be significantly increased by the incorporation of an electric sector, also sometimes termed an electrostatic analyser (ESA). This is composed of a pair of curved plates held at opposite electrical charge, which results in an electric field E. Such a device will transmit ions of all m/z values accelerated by an appropriate potential V. In an

Duty cycle is the proportion of time (%) that a system is active within a given period. In the context of mass spectrometry, this relates to the fraction of the scanned m/z scale that is being measured at a given point in time. If a mass analyser scans over a wide m/z range, but transmits only a single m/z value at each point in time, then many ions are lost, and it will have a low duty cycle.

The kinetic energy of ions possessing charge ze (where z is the number of charges and e the elemental charge), mass m, and velocity v accelerated by potential V is:

$$zeV = \frac{mv^2}{2} \tag{3.1}$$

When passing through a magnetic field B the ions travel with a curved trajectory of radius r, causing them to experience a force given by

$$Bzev = \frac{mv^2}{r} \tag{3.2}$$

Therefore

$$\frac{m}{z} = \frac{B^2 r^2 e}{2V} \tag{3.3}$$

Equation 3.3 describes the relationship between m/z and B, V, and r in a magnetic sector mass spectrometer.

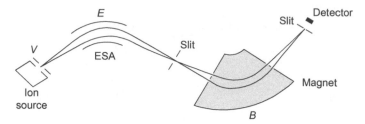

Figure 3.2 Geometry of a double focusing sector mass spectrometer, showing the combination of an electrostatic analyser (ESA) and a magnet.

ESA, ions with similar kinetic energies are focused, whilst those with different kinetic energies are dispersed. Since one of the major factors limiting the resolving power of a mass spectrometer is the spread in kinetic energy of ions possessing the same theoretical m/z, the incorporation of an electric sector greatly improves resolution. The resulting instrument, termed a double-focusing analyser, may incorporate the ESA before or after the magnet. The former arrangement is shown in Figure 3.2. The addition of adjustable slits allows the resolving power of the instrument to be tuned. Narrowing the slits permits high-resolution measurements, but at the cost of lower ion transmission (and therefore lower sensitivity). Spectra with a resolution of 10,000–50,000 (10% valley) can be recorded on such instruments (see Chapter 4 for details on the definition of resolution).

Double focusing sector mass spectrometers were for many years 'the instrument' for high-resolution, accurate mass measurement (see Chapter 4), but with the development of alternative analysers, which offer high resolving powers at full sensitivity, their use has declined somewhat.

Ions accelerated by potential V, travel through an electric field E with a path of radius R, where:

$$E = \frac{2V}{R} \qquad (3.4)$$

3.3 Quadrupole analysers

Single quadrupole instruments are among the most compact, robust, and cost-effective mass spectrometers available. These properties make quadrupoles very popular for many applications, especially in combination with gas and liquid chromatography (see Chapter 7). The analyser is comprised of four parallel metal rods connected, pairwise, to a combination of radio frequency (RF= $V\cos$ (ωt), where V is voltage, ω is angular frequency, t is time), and DC (U) voltages (Figure 3.3). Ions travelling through the quadrupole along the z-axis experience a resulting electric field, and possess stable trajectories providing their position in the xy plane does not exceed r_0 (Figure 3.3). The motion of ions in a quadrupole field is described by equations 3.5–3.7. It can be seen from equation 3.7 that the path of ions in the z direction is independent of any electric potential applied to the quadrupole. Equations 3.5 and 3.6, however, show a clear relationship between motion in the x and y directions, m/z, and the applied RF and DC voltages.

The equations of motion for ions of mass m and charge (ze) travelling through a quadrupole with applied DC voltage U and RF $V\cos$ (ωt).

$$\frac{dx^2}{dt^2} = -\left(\frac{(ze)}{m}\right)\frac{[U+V\cos(\omega t)]}{r_0^2}x, \qquad (3.5)$$

$$\frac{dy^2}{dt^2} = \left(\frac{(ze)}{m}\right)\frac{[U+V\cos(\omega t)]}{r_0^2}y, \qquad (3.6)$$

$$\frac{dz^2}{dt^2} = 0 \qquad (3.7)$$

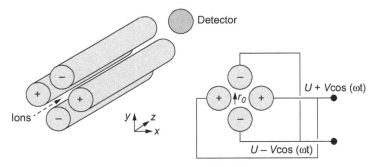

Figure 3.3 Layout of a quadrupole mass analyser, showing the arrangement of the four rods and their electrical connections to the applied RF and DC voltages.

The solutions to differential equations 3.5 and 3.6 are most commonly presented as stability diagrams, which show graphically the combinations of RF and DC voltages that result in stable trajectories for ions of given m/z values (Figure 3.4a). A mass spectrum can be produced by scanning the RF and DC voltages applied to the quadrupole rods. If the two voltages are scanned using scan line 1 (Figure 3.4a), then m/z_1, m/z_2, and m/z_3 possess stable trajectories in the region of intersection between the scan line and stability plot for each m/z value. In other words, these ions would maintain a stable trajectory through the quadrupole to reach the detector. Examination of Figure 3.4a reveals that these regions are unique to each m/z value, and that there is no overlap. This means that the quadrupole is capable of resolving these three species. If the steeper scan line 2 is used, then the regions of stability are reduced significantly (especially at higher m/z), and the resolving power of the quadrupole is increased. The effect of this change on the resulting mass spectrum is shown in Figure 3.4b. Notice that this increase in resolution reduces the number of ions passing through the quadrupole, and hence has a negative impact upon sensitivity. For this reason, quadrupoles are usually operated to achieve unit mass resolution to enable optimum sensitivity. It is also apparent from Figure 3.4a that if the DC voltage U is set to zero, and only the RF voltage is applied, then all m/z values are transmitted by the quadrupole. The result is known as a quadrupole ion guide, and is sometimes used to help focus and transmit ions in a larger mass spectrometer (see Chapter 5). The relatively short path length of ions in a quadrupole mass spectrometer, together with the RF focusing effects of the analyser, means that only modest acceleration voltages (a few hundred volts) and vacuum conditions (10^{-5} mbar) are required for efficient operation. These properties enable quadrupole mass analysers to be incorporated into compact mass spectrometers, which are ideal for use as detectors for gas and liquid chromatographs in a wide range of laboratory settings. As with sector instruments, the quadrupole is normally scanned at rates of the order of 1 scan s^{-1} and a duty cycle of 0.1% over an m/z 1000 range. If the quadrupole is set to transmit ions of a single m/z value, then a duty cycle of 100% is

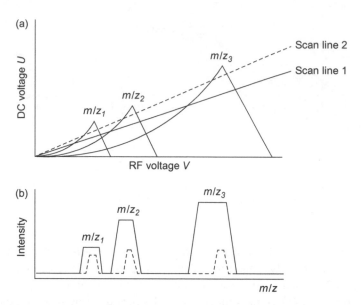

Figure 3.4 (a) Stability diagrams of three ions m/z_1, m/z_2, and m/z_3 in a quadrupole mass analyser. (b) The effect on resolving power of scanning the RF and DC voltages along scan line 1 (solid trace) or scan line 2 (dashed trace).

achieved with a significant gain in sensitivity. The practical upper m/z limit for a quadrupole MS depends upon the RF fields used, and is approximately 2000–4000.

3.4 Time-of-flight analysers

Time-of-flight (TOF) mass analysers have become the mainstay of many mass spectrometry facilities. This popularity is due to their ability to deliver high-resolution, accurate mass measurement reliably and at relatively low cost. The principle of mass analysis in a TOF instrument is straightforward. As shown by equation 3.8 & 3.9, when ions are accelerated by a voltage V, the time they take to traverse a fixed distance in the mass spectrometer flight tube is a function of m/z. Rather than using a continuous ion beam, it is necessary to pulse discrete packets of ions along the flight tube towards the detector in order that their time of flight can be recorded. This can be achieved by pulsing the source extraction voltage (often used in MALDI-TOF instruments), or using an orthogonal pusher (Figure 3.5) to convert a continuous ion beam into an interrupted 'packet' of ions (usually employed in ESI-TOF spectrometers). Both methods require voltages in the kV range to be applied in order to give the ions sufficient kinetic energy to travel along the flight tube to the detector. Flight times are of the order of μs, meaning that scan times can be very rapid, and the typical duty cycle for an orthogonal accelerator-TOF (oa-TOF) is usually of the order of 20%, which is a significant improvement on sectors and quadrupoles operating in full scan mode. Monitoring ions of a single *m/z* value on a standard oa-TOF MS does not bring the same enhancement in duty cycle seen for these other analysers. This is because of the necessity to push packets of ions along the flight tube at discrete intervals.

Ions of a particular *m/z* value inevitably possess a spread of translational kinetic energies (KE), and this limits the achievable resolution. The resolving power of simple *linear* TOF analysers is relatively modest (typically Δ(*m/z*) 0.5 to 1 at *m/z* 1000), but this can be significantly improved by incorporation of an ion mirror, known as a reflectron. The device is composed of a stack of ring electrodes held at decreasingly attractive potentials. Ions entering the rings are decelerated and reflected back with a V-shaped trajectory. Ions of a given *m/z*, with slightly higher KE, penetrate further into the reflectron than those with a slightly lower KE. The result is that the ions are focused to arrive at the detector closer together, hence the *m/z* peak is noticeably narrowed, which provides higher resolution. Figure 3.5 shows the layout of an orthogonal acceleration reflectron-TOF ((oa)re-TOF) analyser. Depending on the length of the flight tube, such instruments are capable of producing data with resolution in the range 10,000–30,000. This property allows properly calibrated TOF type mass spectrometers to deliver accurate mass measurement suitable for determination of elemental composition (see Chapter 4), and resolution of multiply charged ions (typically up to 5$^+$ to 6$^+$) required in peptide analysis (see Chapter 8). A relatively high vacuum (10^{-7} mbar) is usually required in the flight tube of a TOF analyser to ensure efficient ion transmission. TOF analysers can be combined with a quadrupole to produce a Q-TOF hybrid mass spectrometer, which has tandem MS (MS/MS) capability (see Chapter 5). The practical upper *m/z* limit for a linear TOF MS is very high (> 200,000), whilst that of a re-TOF is approximately 30,000.

Ions of mass *m*, charge *ze*, and velocity *v* accelerated by voltage *V* possess energy *E* given by

$$E = zeV = \frac{mv^2}{2} \tag{3.8}$$

Substituting *v* = *d*/*t*, where *t* is time of flight to cover path length *d*, and rearranging for *t*, gives

$$t = \frac{d}{\sqrt{2V}} \sqrt{\frac{m}{ze}} \tag{3.9}$$

Since, for a particular set of instrument parameters, *d*, *V*, and *e* are constants, time of flight is a direct function of *m/z*.

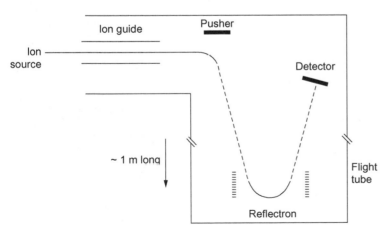

Figure 3.5 Layout of a time-of-flight (TOF) mass analyser that includes a pusher for orthogonal acceleration and a reflectron for ion focusing.

3.5 Ion-trap analysers

The mass analysers considered so far have all been of the 'beam' type, where a continuous or interrupted beam of ions passes through the spectrometer to the detector. We will now turn our attention to trapping mass analysers, which hold ions in a discrete region of space, and measure m/z by interrogating the properties of the confined ions. The first analyser of this type that we will consider is the ion trap, or to give it its full name, the quadrupole ion trap (QIT). As its name suggests, the device shares many similarities with the quadrupole analyser described in Section 3.3. Rather than a beam of ions being allowed to pass through four parallel rods, however, in a QIT the ions are trapped using the quadrupolar electric field. Two basic designs are utilized: the three-dimensional ion trap (also sometimes called the Paul trap), and the linear, or two-dimensional, ion trap (Figure 3.6). The three-dimensional ion trap possesses two dome-shaped end cap electrodes located on either side of a central ring electrode, whereas the linear variant has four rods with a set of end caps. In both cases, low-pressure helium is usually introduced into the trap to provide a buffer gas to cool the ions. The linear trap has the benefit of higher ion capacity, and can more easily be incorporated into tandem or hybrid instruments (see Chapter 5). Despite these differences in design, the principles on which the two types of ion-trap analyser measure m/z are similar.

In a three-dimensional trap, ions are held in the centre of the trap by a combination of RF and DC fields applied to the ring electrode and two end caps. These cause the ions to be accelerated in the axial and radial directions, which results in complex motion, and a stretching and squashing of the ion cloud. Certain combinations of applied fields result in stable ion trapping, whereas other

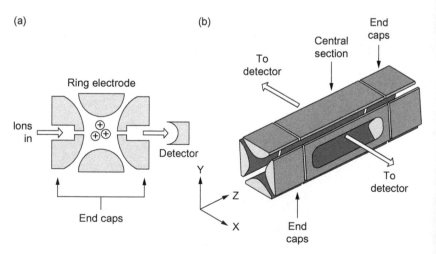

Figure 3.6 (a) Cross-section through a three-dimensional ion-trap analyser. (b) Sketch of a linear ion-trap analyser.

combinations cause the ions to be ejected from the trap. Just as with the quadrupole mass analyser described in Section 3.3, stability diagrams for ions of particular m/z values can be constructed (indeed, the diagrams look rather similar to those for a quadrupole; see Figure 3.4a). If instability is initiated in the z direction by steadily increasing the RF voltage on the ring electrode, ions are sequentially ejected, in an m/z-dependent manner, through perforations in the end cap and subsequently onto a detector. Alternative methods for recording m/z using an ion trap include *resonance ejection*. This uses a supplemental AC voltage applied to the end caps, with a steady increase in the applied RF. The rise in RF results in an m/z-dependent increase in frequency of ion motion within the trap, until it reaches that of the supplemental AC voltage. At this point, resonance is achieved and the ion is activated and ejected from the trap.

In a linear ion trap, the RF and DC fields are applied to the four rods, providing confinement in the radial direction only. Axial trapping is provided by a simple repulsive voltage on the end caps. Scanning the RF and DC voltages on the rods causes the ions to be radially ejected through large gaps between the rods and onto a pair of detectors. By linking detector response with applied voltages applied to the trap, both three-dimensional and linear ion trap designs allow a mass spectrum to be recorded.

In general operation, the resolving power of an ion trap is similar to that of a quadrupole, but significantly higher-resolution measurement is possible if narrow m/z ranges are scanned slowly. Like quadrupoles, ion traps employ low-acceleration voltages, and have relatively modest vacuum requirements (10^{-3} mbar), which allows a compact design. Ion traps can be scanned much faster than quadrupoles, leading to a duty cycle of the order of 10–30% in full scan mode, which is 1–2 orders of magnitude higher than a quadrupole. Quadrupoles do, however, outperform the ion trap if ions of a single m/z value are selected. An additional feature of the ion trap is its ability to perform tandem MS (MS/MS) measurements without the need for a second analyser. This is possible because a precursor ion can be selected, activated, and produce ions measured all within the trap (see Chapter 5). The practical upper m/z limit for an ion trap MS is approximately 2000–4000.

3.6 Fourier transform ion cyclotron resonance analysers

Fourier transform ion cyclotron resonance (FTICR) is another example of an ion-trapping MS technique. The phenomenon of ion cyclotron resonance is based on the absorption of RF radiation at a frequency resonant with the cyclotron motion of ions trapped in a magnetic field. As we have seen in Section 3.2, ions can be deflected by applied magnetic fields. If the momentum of the ions is low, and/or the magnetic field strength is high, they may be 'captured' by the field, rather than just deflected, and made to precess around the field lines (Figure 3.7). The cyclotron frequency f_c of this precessive motion is linked to m/z by equation 3.10. Typical frequencies are in the kHz–MHz range, and thus correspond to the RF region of the electromagnetic spectrum.

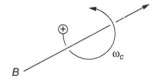

Figure 3.7 Precessive motion of an ion around a magnetic field line.

Ions of mass m and charge ze precess around lines of magnetic field B with cyclotron frequency f_c given by

$$f_c = \frac{\omega_c}{2\pi} = \frac{zeB}{2\pi m} \tag{3.10}$$

where ω_c is the angular cyclotron frequency.

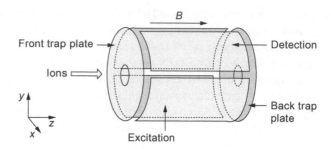

Figure 3.8 Sketch of the FTICR cell showing the pairs of trapping, excitation, and detection plates.

The ICR process is both induced and detected in an analyser cell held in the centre of a homogeneous magnetic field (usually supplied by a superconducting magnet in the 4.7 tesla to 12 tesla range). The cell is composed of three pairs of opposing plates: two excitation plates, through which the RF radiation is applied, two detector plates, and the front and back trapping plates (Figure 3.8). Ions are radially trapped in the x, y plane by the applied magnetic field, and in the axial z direction by a repulsive potential on the trapping plates. To allow ion injection into the cell, the voltage on the front plate can be momentarily lowered. Once in the cell, ions are excited by application of an RF pulse, which—if resonant with the cyclotron frequency ($f = \omega_c/2\pi$)—has the dual effects of (i) causing the ion orbit to become coherent with that frequency and (ii) increasing the cyclotron radius of the ions. This increase in radius brings the ions in proximity to the detector plates, and an image current of their orbits can be recorded. In practice, rather than scanning the RF frequency and detecting resonance of each m/z value sequentially, a broadband RF pulse, or 'chirp', is applied, and all ions are brought into resonance at once. Following excitation, the ions are allowed to relax to their original cyclotron radii, which results in the detection of a transient signal (Figure 3.9). Fourier transformation of the time-domain transient gives the individual frequencies f, which are related to m/z by equation 3.10. Thus, a mass spectrum of the ions is recorded.

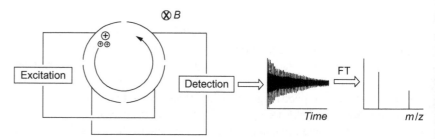

Figure 3.9 Detection of the transient, or free induction decay (FID), signal of ions in an FTICR cell, and subsequent Fourier transform (FT) of the time domain signal to its composite frequencies, and hence m/z of each ion (by equation 3.10).

The resolution achieved by FTICR MS measurement depends upon the transient time recorded, and also falls off linearly with m/z. If the cyclotron radii of the ions are allowed to decay slowly (over 10 seconds), then extremely high-resolution measurement is possible ($R = 1,000,000$). In practice, a duty cycle time in the order of 1 second is required for many applications, but even at this data acquisition rate, a resolution of 100,000 is still possible. This property makes FTICR MS an extremely powerful technique in both small and large molecule analysis. An ultra-high vacuum (10^{-10} mbar) is required in the FTICR cell to reduce ion collisions with air molecules, and thereby facilitate extended measurement of the transient. This means that several large turbomolecular pumps are required to maintain the vacuum which, together with the need for a superconducting magnet, makes FTICR instruments large and expensive. To achieve ion injection into the cell it is necessary to penetrate the fringe-field of the magnet. This is achieved either by application of a high voltage to accelerate ions through the field, or by use of an RF-only multipole to keep the ions focused. In both cases, ions are usually pulsed into the cell in packets with a synchronized drop in the voltage on the front trapping plate. The duty cycle of an FTICR instrument is improved dramatically by the addition of a trapping hexapole or octupole ion guide at the front of the instrument. These RF-only multipoles focus ions radially (see Section 3.3) and possess switchable trapping electrodes at the entrance and exit. Ions entering from the source can be accumulated before being passed on to the transfer optics leading to the FTICR cell. If the operation of the trapping octupole and FTICR cell are synchronized, duty cycles approaching 100% can be achieved. In practice, such a high duty cycle is only possible for scan times of approximately 1 s or less, due to the limited ion storage capacity of the octupole and the FTICR cell itself. The practical upper m/z limit for an FTICR MS depends upon the size of the magnet, but can extend above 10,000.

It should be pointed out that the motion of ions in an FTICR cell is more complex than indicated in Figure 3.9, being composed of cyclotron motion (as described), trapping motion (the oscillation between trapping plates), and magnetron motion (a slow rotation of the centre of the cyclotron motion about the central axis of the cell). In practice, however, we need only concern ourselves with cyclotron motion in order to understand how FTICR records a mass spectrum.

It is possible to perform MS/MS in an FTICR cell by using a special precursor ion selection RF waveform that ejects all ions from the cell except the desired ion. This precursor can then be activated in the cell in a number of ways, including collision-induced dissociation (CID), laser photodissociation, and electron-capture dissociation (see Chapter 5). FTICR analysers may also be incorporated into hybrid mass spectrometers, including a front-end quadrupole and collision cell, or an ion-trap analyser.

3.7 Orbitrap analysers

The orbitrap is the most recent design of analyser to find widespread use in commercial mass spectrometers. It consists of a pair of elongated cup-shaped outer electrodes containing a central inner electrode, or spindle. Ions injected into the

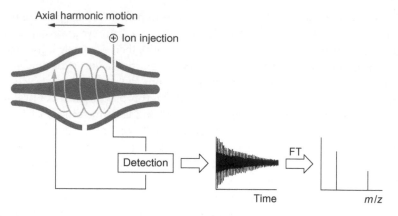

Figure 3.10 Detection of axial harmonic motion of ions in an orbitrap analyser and conversion of the transient FID signal, in the time domain, to its composite frequencies (and hence *m/z*) by Fourier transformation.

Ions in the orbitrap experience a field which causes them to oscillate in the axial direction with a frequency ω_{orb} which is related to *m/z* by:

$$\omega_{orb} = \left(\left(\frac{ze}{m} \right) k \right)^{\frac{1}{2}} \qquad (3.11)$$

where *k* is field curvature.

evacuated space between the outer and inner electrodes orbit the spindle due to a balance of electrostatic and centrifugal forces (Figure 3.10). To facilitate ion injection the voltage on the spindle is first reduced, and then ramped up to push the ion cloud to the desired orbit around the spindle (termed electrodynamic squeezing). In orbitrap mass analysers, the relationship between axial harmonic motion of the ions (controlled by the geometry of the orbitrap) and their *m/z* is exploited for mass measurement.

Motion is detected by an image current generated as the ions oscillate axially between the outer electrodes whilst maintaining an orbital motion around the spindle. Fourier transform of the composite signal allows extraction of individual frequencies ω_{orb}, and therefore individual *m/z* values (by equation 3.11). As with FTICR, the frequency of this motion is in the kHz–MHz range. The more oscillations detected the higher the resolving power, with data exhibiting resolution in excess of 200,000. This high-resolution capability has made the orbitrap an attractive mass analyser. Its performance is surpassed only by FTICR MS, and comes without the need for a large and expensive superconducting magnet. Moreover, the resolving power of an orbitrap decreases with the square root of the ion's *m/z*, which is preferable to FTICR, where a linear decrease is observed. Orbitraps, like FTICR instruments, do however require an ultra-high vacuum (10^{-10} mbar) in order to reduce ion–molecule collision in the trap and achieve high resolution.

One of the challenges associated with orbitrap analysers is efficient ion injection into the trap. In current commercial instruments this is solved using a 'C-trap', which collects and bunches ions for injection into the orbitrap. This has the added benefit of increasing the duty cycle, as the C-trap can be synchronized to fill with ions whilst the orbitrap detects. For example, spectra with a resolution of 17,500 can be obtained at rates of approximately 10 scans s^{-1}, with a duty cycle approaching 100%. Longer detection times (required for higher resolution) reduce the duty cycle accordingly. The practical upper *m/z* limit for an orbitrap is impressively high for such a compact analyser, at over 10,000.

Table 3.2 Properties of mass analyser types.

Analyser	Properties[a]				
	Typical m/z limit	Resolution achievable[b]	Scan time (s)	Duty cycle (%)[c]	Operating pressure (mbar)
Magnetic sector	5,000	100,000	0.5–2	0.1–2	10^{-7}
Quadrupole	4,000	2,000	0.1–1	0.1–10	10^{-5}
Linear TOF	250,000	5,000	100 μs	≈ 100	10^{-7}
Reflectron TOF	30,000	30,000	100 μs	5–20	10^{-7}
Ion trap	4,000	2,000	0.01–1	10–30	10^{-3}
Ion cyclotron resonance	20,000	1,000,000	0.5–10	≈ 90[d]	10^{-10}
Orbitrap	20,000	250,000	0.1–1	≈ 90[d]	10^{-10}

[a] Properties quoted are those typical of commercially available instruments under normal operating conditions. Specially modified spectrometers/operating conditions may give rise to enhanced performance. [b] FWHM definition (see Section 4.2 for details). [c] In full scanning mode. [d] Providing ions are accumulated in a synchronized trap.

MS/MS is not normally performed inside the orbitrap analyser itself. In most commercial instruments, this functionality is provided by including a front-end quadrupole and collision cell, or an ion-trap analyser to produce a hybrid spectrometer.

3.8 Summary

A comparison of some of the properties of the major mass analyser types is provided in Table 3.2. From the material presented in this chapter you should be familiar with the basic physical principles behind the main types of mass analysers used in contemporary mass spectrometry.

You should be able to explain the operating principles of the following:

- Magnetic sector (including double focusing instruments).
- Quadrupole.
- Time-of-flight analyser.
- Ion trap.
- Fourier transform ion cyclotron resonance analyser.
- Orbitrap analyser.

You should also be aware of the performance capabilities of each type of analyser and be able to compare their advantages and disadvantages.

3.9 **Exercises**

3.1   A magnetic sector mass analyser has radius $r = 1.00$ m and operates at an acceleration voltage $V = 5.00$ kV. If the maximum operating magnetic field $B = 1.00$ T, calculate the upper limit of m/z measureable on the instrument (m should be expressed in Da, where 1 Da $= 1.66 \times 10^{-27}$ kg. The elemental charge $e = 1.60 \times 10^{-19}$ C).

3.2   Without redesigning the instrument in Question 3.1, what simple change could you make to its operating parameters in order to extend the m/z range? Can you identify any disadvantages that may result from this change?

3.3   Calculate the time of flight t for an ion of m/z 1000, accelerated by a voltage $V = 20.0$ kV, to travel along a flight tube of 1.00 m. Recall that 1 Da $= 1.66 \times 10^{-27}$ kg, and the elemental charge $e = 1.60 \times 10^{-19}$ C.

3.4   What is the role of the reflectron in a TOF analyser? Does the use of a reflectron have any disadvantages?

3.5   What is the main advantage of high-resolution measurement on a TOF compared to a double focusing sector analyser?

3.6   List three similarities and three differences between quadrupole and ion trap mass analysers.

3.7   What features of a quadrupole mass analyser make it an ideal 'detector' for use with gas or liquid chromatography?

3.8   Calculate the applied RF frequency (in Hertz) required to bring an ion of m/z 500 into cyclotron resonance in a magnetic field $B = 9.40$ T. Recall that 1 Da $= 1.66 \times 10^{-27}$ kg, and the elemental charge $e = 1.60 \times 10^{-19}$ C.

3.9   What are the principal requirements for achieving high-resolution mass measurement on FTMS instruments?

3.10 What is the principal advantage of an orbitrap over an FTICR analyser?

3.10 **Further reading**

Chapman, J. R. (1993). *Practical Organic Mass Spectrometry*, 2nd edn. Chichester: Wiley.

Cole, R. B. (ed.) (2010). *Electrospray and MALDI Mass Spectrometry*. New York: Wiley.

Watson, J. T. and Sparkman, O. D. (2007). *Introduction to Mass Spectrometry*, 4th edn. Chichester: Wiley.

4 Resolution, accurate mass, and sensitivity

4.1 Introduction

Resolving power, mass accuracy, and sensitivity are three key properties of a mass spectrometer. Put simply, these are (i) the limit to which an instrument can separate ions of different m/z value (see Section 4.2 for clarification of resolving power vs resolution), (ii) the error limit within which an instrument can measure an ion's m/z value relative to its theoretical (exact mass) value, and (iii) the relationship between the ion current measured by the mass spectrometer and the concentration of the sample analyte. The term 'sensitivity' is sometimes used synonymously with limit of detection; that is, the lower limit of sample required to elicit an acceptable signal response.

It is worth pointing out that the terms 'high resolution' and 'accurate mass' can be a source of confusion to some outside the field of mass spectrometry, and are sometimes incorrectly used interchangeably. Given that data from accurate mass measurements are commonly used in the characterization of synthetic products, it is important to clarify these distinct concepts. Similarly the contrast between 'sensitivity' and 'limit of detection' is not always made clear in the literature, and the terms are often used synonymously despite having different meanings.

The aims of this chapter are to explain resolving power, resolution, accurate mass, and sensitivity, and to show how these properties can be used for analytical purposes.

4.2 Resolving power and resolution

Resolving power is used to describe the separating capability of a mass spectrometer or mass analyser (i.e. the ability to deliver a specified level of mass resolution). It is usually expressed as a difference in separable m/z values ($\Delta(m/z)$) at a given value of m/z. *Resolution* (R), in contrast, refers to the separation obtained in a spectrum (i.e. resolution is a property of the data rather

Resolution observed in a mass spectrum is given by:

$$R = \frac{m/z}{\Delta(m/z)} \qquad (4.1)$$

where $\Delta(m/z)$ is either the spectral peak width or difference in m/z values that can be separated at a defined value of m/z. Thus, if a mass analyser is capable of separating m/z 1001 from m/z 1000, it has a resolving power of 1 at m/z 1000, and the resulting spectrum would exhibit a resolution of $1000/1 = 1000$.

The resolution and resolving power controversy. There is some debate as to whether equation 4.1 should be used to express resolution or resolving power, with some textbooks preferring to use $\Delta(m/z)$ to denote resolution instead. The current IUPAC definition, however, employs equation 4.1 to mean resolution, as does the majority of the mass spectrometry community. For those reasons, resolution is given by equation 4.1 here.

than the analyser), and is calculated by equation 4.1. The equation includes the variable m/z, as well as $\Delta(m/z)$. Thus, it takes into account the specific m/z value at which the mass separation, or peak width, $\Delta(m/z)$ is achievable. This is an important consideration, as some types of mass analyser (e.g. quadrupoles) are often operated to deliver a constant unit mass resolution, regardless of m/z, whereas with others (e.g. TOF and FTMS analysers), $\Delta(m/z)$ varies significantly with m/z.

In the previous definitions we have referred to separation of m/z values and peak width without actually saying what we mean by these terms. When do we consider that two ions are sufficiently separated on the m/z scale, and how do we measure peak width?

Separation is usually defined in terms of the difference between the maxima of two peaks of equal intensity, allowing for a specified degree of overlap. Figure 4.1a shows m/z values 1000 and 1001 with a peak overlap at 10% of the signal intensity. This is the so-called 10% valley definition, and we would say that the spectrum exhibits a resolution of 1000 (remembering that $R = m/z/\Delta(m/z)$ = 1000/1). For historical reasons this definition of resolution is usually reserved for magnetic sector mass analysers, and it has the obvious disadvantage that it requires the detection of two ions of equal intensity in order to satisfy the strict definition.

An alternative method for reporting spectral resolution is to use peak width as the $\Delta(m/z)$ value. By convention, this is measured at half peak height and is referred to as the full width at half maximum (FWHM) definition of resolution. Figure 4.1b shows an ion of m/z 1000 with a FWHM $\Delta(m/z)$ of 0.5. Comparison of this spectrum with that in Figure 4.1a shows two interesting features. First, it is not necessary to have a pair of ions to determine resolution using the FWHM definition. Secondly, despite identical peak widths in the two spectra (a) and (b), the resolution, calculated using FWHM is 2000—twice that of the 10% valley value. This difference arises because $\Delta(m/z)_{FWHM} = 0.5$, whilst $\Delta(m/z)_{10\%valley} = 1$. Thus, when stating the resolution obtained, or the resolving power of an analyser, it is important to specify which definition is used. In practice, FWHM is normally used for all non-sector mass analysers.

Having considered the definitions of resolving power and resolution, we can now examine the analytical advantages afforded by high-resolution measurement. Consider the spectra shown in Figure 4.2a–c. Each spectrum contains the same three ions, which have very similar m/z values, but which are recorded with different resolving powers. It is easy to see that with a resolution of 2000 it is not possible to separate the three ions, and essentially only one peak would be detected. With an intermediate resolution of 20,000, two of the three ions would be resolved, but only in the highest-resolution spectrum are all three ions separated as distinct species. Clearly, if these three ions were to be present in a sample, the highest-resolution spectrum would be essential to analyse them individually. Thus, for complex samples containing

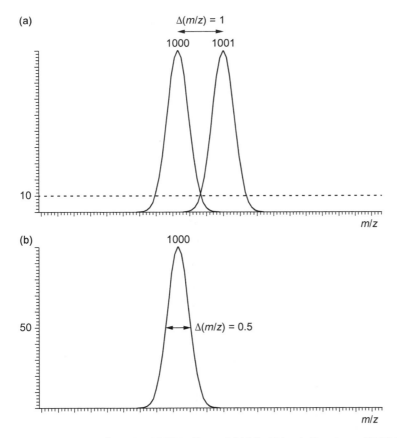

Figure 4.1 Mass spectra illustrating (a) 10% valley, and (b) full width at half maximum (FWHM) definitions of resolution.

many ions with similar m/z values, high-resolution measurement is a distinct advantage.

Analysers with high resolving power are also of significant advantage in measuring large molecules, such as peptides and proteins. Electrospray ionization (ESI) typically generates multiply-charged ions for peptides above approximately 1 kDa in mass. As the charge (z) increases, the gap between neighbouring isotopes decreases on the m/z scale, since Δm remains 1, but z is > 1. Thus, the isotopic peaks of a doubly-charged ion are separated by $\Delta(m/z)$ 0.5 (see Figure 1.7). A consequence of this is that the resolving power of the spectrometer needs to be twice that required to separate a singly charged ion at the same value of m/z (remember we need to specify the m/z when quoting resolving power).

As z becomes larger, many analysers are unable to achieve isotopic resolution, and it becomes necessary to change from using monoisotopic to average mass.

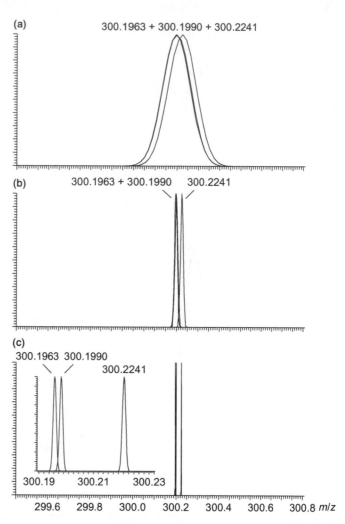

Figure 4.2 Mass spectra illustrating FWHM resolution values of (a) 2000, (b) 20,000, and (c) 200,000. Note that the highest resolution is required to separate all three ions. Intensity is shown on the vertical axis.

4.3 Accurate mass measurement

Clearly, when recording a mass spectrum we would want the measured *m/z* values to be both accurate and precise at all times, but the phrase 'accurate mass measurement' has a particular meaning in mass spectrometry. It is usually taken to denote a degree of accuracy that allows determination of the elemental composition of an ion. This normally means an error limit in the order of a few ppm, which for ions of a few hundred daltons requires measurement to 3 or 4 decimal places. If the ion measured accurately is the molecular ion, then the elemental composition of the molecule is obtained. This is usually considered as an important component in the characterization of novel synthetic

Accuracy and *precision*. Accuracy refers to how close a measured value is to an expected theoretical one, whereas precision relates to the reproducibility of a measurement. Thus, accuracy without precision is of very limited value.

In mass spectrometry the error in mass accuracy is normally reported in parts per million (ppm) = [(measured mass – exact mass)/exact mass] × 1 × 10^6.

products. Thus, accurate mass measurement is of general importance in synthetic chemistry.

As mentioned at the beginning of this chapter, the terms 'accurate mass' and 'high resolution' are often used interchangeably, but this is incorrect. In principle, high-resolution measurement is not always required for accurate mass measurement, but it is a distinct advantage, and is very important in some circumstances. For example, consider the spectra in Figure 4.2. If we were attempting to measure accurately the masses of these ions, we could only do so if they were well resolved, as in Figure 4.2c. In the low-resolution spectrum of Figure 4.2a, a mass corresponding to the weighted average of all three individual masses would be recorded, and clearly this would compromise the accuracy obtained. Where no such overlap exists, the need for high resolution is not essential. In practice, however, accurate mass measurement is often performed using mass analysers capable of producing spectra with a resolution in excess of 10,000 (FWHM). As outlined in Chapter 3, this is possible with sector, TOF, orbitrap, and FTICR analysers.

The most important factor in obtaining an accurate mass spectrum is high-quality calibration of the m/z scale. A number of calibration methods can be employed, as follows.

4.3.1 Internal calibration

An internal calibration is usually considered as the 'gold standard' method of calibration for accurate mass measurement and tends to produce the most reliable results. We will use internal calibration to illustrate the overall process of accurate mass measurement. Essentially, in this approach, the mass spectra of the calibrant and sample are recorded simultaneously. Ideally the internal calibrant will produce a number of ions spread across the mass range of interest. Providing that the elemental composition of each calibrant ion is known, then its exact mass is also known, and this can be used to correct the measured m/z scale of each individual scan recorded. Internal calibration, therefore, results in both high-precision and high-accuracy mass measurements across a defined mass range. Figure 4.3 shows the EI mass spectrum of hexabromobenzene acquired with the commonly used internal calibrant perfluorotributylamine (PFTBA). Perfluorinated compounds are widely used for calibrating EI spectra, as the fluorine atoms add significant mass (to cover a wide m/z range) without complicating the isotopic pattern or decreasing volatility of the calibrant. Application of a calibration to the spectrum, using the PFTBA ions, results in a correction of the mass scale across the m/z range, and the production of accurate mass data for the hexabromobenzene-derived ions. The insert shows an expansion of the region around the molecular ion of hexabromobenzene. Its complexity is largely due to the presence of six Br atoms in the molecule, which exist as a statistical combination of ^{79}Br and ^{81}Br isotopes. As described in Chapter 1, where we have isotopic resolution, by convention we take the monoisotopic mass composed of the most abundant isotopes of each element. In this case it is $^{12}C_6{}^{79}Br_6$, which has an exact (theoretical) mass of 545.5100 (derived from

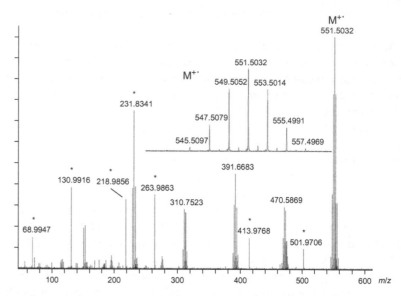

Figure 4.3 EI mass spectrum of hexabromobenzene measured with internal calibration using perfluorotributylamine (calibrant ions are marked *). The insert shows details of the complex molecular ion, which arises due to the natural abundance of ^{79}Br and ^{81}Br isotopes. The expected m/z for $^{12}C_6{}^{79}Br_6$ is 545.5100, corresponding to an error in mass measurement of 0.4 ppm.

The importance of the mass of the electron. When measuring the accurate mass or, more strictly, m/z of an odd electron ion (i.e. a radical cation or anion) it is important to take the mass of the electron into account. On the Dalton scale, an electron has a mass of 0.55 mDa, which can easily affect the third decimal place of a mass measurement. Thus, for comparison with positive ion EI spectra, 0.55 mDa should be subtracted from the exact (theoretical) mass of the neutral molecule. With even electron ions (e.g. cations without an unpaired electron), commonly seen as fragments in EI spectra, the mass of the electron should not be subtracted. For ionization methods that result in protonation, or deprotonation, of a molecule it should be ensured that the mass of the proton is added or subtracted, respectively, and not that of a hydrogen atom (proton + electron). Some software packages for interpreting accurate mass data take the electronic mass into account; but not all do.

the values in Table 4.1) as a neutral molecule, but a value of 545.5095 as a radical cation $[^{12}C_6{}^{79}Br_6]^{+\bullet}$, which is relevant for comparison with a mass measured by EI ionization. As can be seen from Figure 4.3, the measured mass (or more correctly m/z) for this ion is 545.5097, representing an error of 0.2 mDa, or 0.4 ppm. A mass accuracy of < 1 ppm is generally considered to be more than sufficient for chemical characterization purposes, but see the discussion in Section 4.3.5.

One drawback of internal calibration is the need to mix sample and calibrant. For instruments using EI ionization this is partially addressed by the use of a separate reservoir for the calibrant compound (e.g. PFTBA) so that it is only mixed with the analyte in the EI source, but in other ionization methods it is often necessary to dope calibrant into the sample prior to introducing it into the spectrometer. This can often lead to signal suppression of one or the other components in the mixture, and care is needed to produce a spectrum with similar intensities of analyte and calibrant ions. A second problem is that addition of a calibrant is often incompatible with coupled chromatographic separation (see Chapter 7).

4.3.2 External calibration

With external calibration, the spectrum of a calibrant is acquired and the instrument is calibrated separately from measurement of the sample. This is considered to be less reliable than internal calibration, as mass correction is not applied simultaneously with sample measurement, which opens up the opportunity for drift in instrumental conditions between the two acquisitions. Such effects can result in significant errors between measured and exact (theoretical) ion masses, which can make determination of elemental composition unreliable.

Providing that instrumental conditions remain constant, it is possible to use external calibration with confidence. This requires either an analyser with an extremely stable calibration of the m/z scale, or measurement of the calibrant immediately before or after sample analysis. FTMS analysers generally offer sufficient stability such that mass accuracy of a few ppm can be achieved even days after calibration. TOF analysers, however, are much more susceptible to environmental conditions, and accurate mass measurement usually requires close external calibration (ideally within a few minutes of sample measurement). This is possible if analysis times are short, as is the case with direct sample injection, but the approach is difficult to combine with chromatography, where run times may be up to an hour or longer (see Chapter 7).

4.3.3 Lock mass correction

This method is a hybrid of internal and external calibration. Full calibration of the mass scale is performed externally, but then a single reference ion is used to adjust calibration of the m/z scale internally as the sample is being measured. If an EI source is used, then the lock mass compound is usually fed into the ionization source using a reservoir separate from the sample probe. With ESI, a second, dedicated lock mass probe is usually employed. The analyte and reference sprays are sampled alternately by a rotating baffle, which allows ions from only one probe to enter the spectrometer at a time. By using a single ion, rather than a full calibration series, there is less chance of sample–calibrant overlap in the spectrum, and in some source designs the lock mass ion is acquired in a separate data channel from that of the analyte, so it does not interfere with the analyte spectrum.

4.3.4 Peak matching

Accurate mass measurement by peak matching is now relatively rarely employed, but is included here for completeness. It is usually carried out on sector field mass spectrometers operating at high resolving power, and makes use of a reference mass of known exact mass. First, a calibrant ion close in m/z to the expected analyte mass is selected, and the magnet is set at a fixed field B to transmit this ion at a given acceleration voltage V_1. Next, the sample is introduced, and the acceleration voltage adjusted until the analyte ion is transmitted at the fixed magnetic field value B. The ratio of this new acceleration voltage (V_2) to V_1 is inversely proportional to the ratio of the two m/z values, and thus the mass of the analyte ion can be determined. To achieve high accuracy, the voltages are rapidly switched between V_1 and the adjustable V_2 whilst the m/z peaks are aligned on a screen. The method can produce very accurate results, with errors routinely < 1 ppm.

Table 4.1 Monoisotopic masses of selected elements.

Element	Monoisotopic mass[a]
H	1.00783
C	12.00000[b]
N	14.00307
O	15.99492
F	18.99840
P	30.97376
S	31.97207
Cl	34.96885
Br	78.91834
I	126.90447

[a] Mass of the most abundant isotope to 5 d.p.

[b] By definition.

4.3.5 **Applications of accurate mass measurement**

As we have already seen in Figure 4.3, with appropriate calibration it is possible to record the mass of a molecule (via its molecular ion) to sub-ppm accuracy. This is often sufficient to confirm the elemental composition of a synthetic product, or even to predict that of an unknown compound; but how is this done, and what level of accuracy do we need to achieve for a valid accurate mass measurement?

Consider the exact monoisotopic masses of the elements shown in Table 4.1. It is apparent that different combinations of elements will have different exact masses, even if the nominal masses are the same. For example, the molecular formulae of propane and ethanal (acetaldehyde) are C_3H_8 and C_2H_4O, respectively. Both have a nominal molecular mass of 44 Da. Their exact molecular masses, however, are different, being 44.0626 and 44.0262 Da (to 4 d.p.), respectively. If we were to analyse these molecules by EI-MS, we would record molecular ions of m/z 44.0621 and 44.0257 (remembering to subtract 0.55 mDa for the mass of the electron). Thus, if we could measure the m/z of each of these ions accurately, we would be able to distinguish between them and assign their formulae.

For the examples of propane and ethanal, it is instructive to examine what degree of mass accuracy is required to have confidence in assignment of their elemental compositions. The difference between the masses of their molecular ions is 44.0621–44.0257 = 0.0364, or 827 ppm. This is much greater than the typical errors of a few ppm associated with accurate mass measurement, so assuming these were the only two possibilities under consideration, it would be very straightforward to identify which of these two compounds was present in the sample. In fact, for small ions such as these, very modest mass accuracies (errors of a few hundred ppm) are often sufficient, providing of course that the measurements are precise (reproducible). The reason for this is that there is a relatively *large* mass difference in the two *low* molecular mass values. Naturally, as the mass of the analyte ions increases, or the difference between possible m/z values decreases, the need for higher mass accuracy (in ppm terms) becomes much more important. For example, comparing the exact masses of the molecular ions $[C_{17}H_{24}N_4O]^{+\bullet}$ m/z 300.1946, and $[C_{16}H_{28}O_5]^{+\bullet}$ m/z 300.1933 gives a difference of 0.0013, which corresponds to a much more demanding 4.3 ppm, meaning that a mass measurement with a reliable error of approximately 1 ppm would be needed to ensure correct assignment.

The effect of increasing mass on the number of possible combinations of elements, and thereby the number of alternative formulae, also dramatically increases the need for accuracy in mass measurement. Figure 4.4 illustrates the relationship between molecular mass and the number of molecular formulae possible within 5 ppm, 1 ppm, and 0.1 ppm mass errors. It is clear from this plot that 5 ppm, which is often taken as an acceptable error limit, only produces unique chemical formulae for masses in the range < 100 Da. By a mass of 500 Da, tens of formulae are possible, and by 1000 Da the number of possibilities is in the order of 150. Moreover, these results are for formulae containing C, H, N, and O only. Clearly, the inclusion of other elements would increase the number of possible answers yet further.

Some of the theoretical formulae included in Figure 4.4 can be discounted on the grounds of low chemical likelihood, and also through properties of the

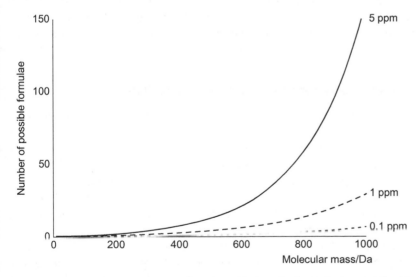

Figure 4.4 Plot of molecular mass vs number of possible formulae (using C, H, N, and O only) within mass accuracy ranges of 0.1, 1, and 5 ppm. Note the exponential relationship and the need for very high mass accuracy to reduce the number of possibilities at higher mass.

mass spectrum other than mass accuracy, such as isotope composition of the molecular ion. Nonetheless, very high mass accuracy (< 1 ppm) with high precision is needed to identify unique elemental composition by MS. In skilled hands, such errors are reproducibly achievable on sector field and FTMS instruments. Through advances in detector and ion optic designs, high-end TOF analysers are also now beginning to approach these levels of accuracy.

In answer to the question posed at the beginning of this section, 'what level of accuracy do we need to achieve for a valid accurate mass measurement?', the response must be that it depends on what we wish to show. If it is that a measured mass is sufficiently close to an exact theoretical value, and is thereby *compatible* with the corresponding formula, then a sub-5 ppm error may be sufficient. But if the intention is to show that a particular measured mass corresponds to a unique formula, then much higher accuracy will almost certainly be required for all but the smallest of molecules. For this reason, it is not recommended to set a fixed acceptable error limit for accurate mass measurement, but to achieve the best value possible under the circumstances and be aware of any resulting limitations this places on the data.

4.4 Sensitivity

Mass spectrometry is well known for its high sensitivity compared to some other spectroscopic techniques such as NMR. For example, detection of compounds at femtomolar levels can be routinely achieved for some analytes. It has been said that, in general, more chemical information can be determined by mass spectrometry, per amount of sample required, than by any other analytical technique. But what is meant by the term 'sensitivity' in the context of mass spectrometry, and what factors affect the sensitivity of a mass spectrometry experiment?

4.4.1 What do we mean by the term 'sensitivity'?

Sensitivity in general is defined as the slope of the curve which relates the signal response to the concentration of an analyte. The steeper the curve, the greater the signal response for the same concentration, and the more sensitive the analysis. In mass spectrometry this translates into the relationship between the ion current detected and the concentration of the analyte in the sample. Figure 4.5 demonstrates this relationship, and it can be seen that the steeper the slope (Analysis 1) the more ions are measured for the same analyte concentration and therefore the more sensitive the analysis.

As we saw in Chapter 2, the number of ions produced is highly dependent on the type of ion source used. Some ionization techniques are optimal for certain types of compounds; for example, polar organic compounds by ESI, or volatile compounds by EI and CI. The sensitivity of a mass spectrometer is therefore not universal and varies depending on the instrumental design and the analyte of interest. The *sensitivity of a particular mass spectrometry experiment*, which takes

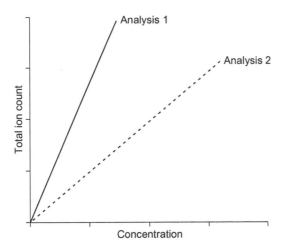

Figure 4.5 The slope of the curve relating ion count and concentration defines the sensitivity of a mass spectrometry experiment. Higher sensitivity is exhibited in Analysis 1 than in Analysis 2.

into consideration the nature of the analyte, rather than just the sensitivity of the mass spectrometer, is often a more useful concept.

4.4.2 Sensitivity and limit of detection

Terms that are often used synonymously with sensitivity are *lower limit of detection* and *lower limit of quantification* (LLOD and LLOQ respectively). LLOD and LLOQ are related and important concepts, but strictly are not equivalent to sensitivity in the context of mass spectrometry. We shall discuss briefly LLOD here, but further discussion of LLOD and LLOQ is also provided in Chapter 7. LLOD refers to the lowest level of an analyte that produces detected ions that are distinguished from background noise (chemical and electronic). The more sensitive the analysis the lower the LLOD can be; however, LLOD is also a function of *selectivity*, which is the ability of the mass spectrometer to isolate a particular analyte in the presence of others. Approaches to improve selectivity can include tandem mass spectrometry (see Chapter 5) and Selected Ion Monitoring (SIM) experiments, as well as tuning the resolution at which the mass spectrometer operates. These approaches all increase selectivity by reducing background noise, leading to a higher signal-to-noise ratio and improved LLOD. However, these approaches do not usually increase the number of analyte ions reaching the detector nor improve absolute sensitivity, and often the number of analyte ions detected is in fact reduced. In such cases, LLOD is improved because the background noise is reduced to a greater extent than the loss of analyte ions. In this case, sensitivity is reduced despite a lower limit of detection being achieved. A complicating factor is that different mass spectrometer configurations (mass analysers and their combinations) enable improvements in LLOD to a greater or lesser extent using these methods. In general, however, tuning parameters which lead to higher resolution can lead to better selectivity, but these are usually at the expense of sensitivity.

4.4.3 **Factors affecting sensitivity**

Most mass spectrometers are extremely inefficient in absolute terms when it comes to creating and transferring ions. The number of ions generated versus how many actually enter the high-vacuum region of the mass spectrometer is low: often < 0.1% of ions produced by EI, CI ESI, and APCI sources are transmitted to the analyser. Ions from an ESI source may be scattered by collisions with neutral gas molecules, or lost by striking the surface of the ion source. In contrast to ESI, nano-ESI is highly efficient. In excess of 30–40% of ions generated can be transmitted as a result of the low flow of analyte in nano-ESI sources and the smaller scale of the system relative to the inlet of the mass spectrometer. This makes nano-ESI one of the most efficient and sensitive ionization techniques available for mass spectrometry.

Once inside the mass spectrometer, ions may have a very different path to travel depending on the type of mass analyser and its configuration. These differences can lead to a very wide range of ion transmission efficiencies, and therefore overall sensitivities, for the analysis of particular compounds.

4.4.4 **The duty cycle and sensitivity**

Ion transmission efficiencies inside the mass spectrometer are significantly affected by the duty cycle of the analyser. As discussed in Chapter 3, 'beam' type analysers such as magnetic sectors, quadrupoles, and TOF systems scan, or otherwise interrupt, the ion beam during analysis. For example, whilst a quadrupole analyser is scanning across a particular mass range, only ions of a specific m/z value are being transmitted through to the detector at any one time, and all other ions are neutralized on the quadrupoles or the inner wall of the mass spectrometer. This leads to the loss of a significant proportion of ions. Of the beam type analysers, TOF systems are perhaps the most efficient at around 20% transmission, whilst others are often not more than 1–2%. Trapping analysers such as ion traps, FTICR, and orbitraps tend to have much more efficient duty cycles, and some, such as FTICR and orbitraps, can approach 100% (see Chapter 3).

4.4.5 **The effect of mass range and tandem mass spectrometry approaches**

Certain mass spectrometry experiments can be used to enhance sensitivity and detection limits for the analysis of selected ions once they have entered the mass spectrometer. The effectiveness of these methods depends on the type of mass analyser used, however. Commonly, mass spectrometry measurements involve acquisition across an m/z range. For 'beam analysers', which scan across the mass range over a specific period of time, reducing the mass range increases the time spent scanning each ion, and therefore more ions will be detected at a particular m/z value. Taken to its logical conclusion if the mass analyser is set to scan only a very narrow mass range (1 m/z unit, for example), the number of ions transmitted will be maximized and the sensitivity of the analysis of that ion will be optimized. This is known as selected ion monitoring (SIM), and, for beam analysers, it usually provides a significant increase in sensitivity, which also leads to lower LLODs. SIM experiments, however, afford much less of an advantage when using TOF analysers, which measure a complete mass range extremely rapidly (μs) and hence are more sensitive across a wide m/z range. Hybrid instruments (combining more than one mass analyser) such as triple quadrupoles and quadrupole-orbitraps can provide additional benefits in SIM mode by being able to select a number of specific m/z values to monitor during each data acquisition cycle. This can simultaneously increase sensitivity and selectivity for multiple analytes of interest in complex sample mixtures. Experiments which can enhance selectivity even further include selected reaction monitoring (SRM) and multiple reaction monitoring (MRM) (see Chapter 5). These work by isolating and fragmenting individual precursor ions. SRM and MRM have the advantage of reducing the background signal significantly, and hence lowering the LLOD. In order to be detected, a compound has to have a particular precursor m/z, and fragment to give a particular product m/z. Sensitivity is technically not enhanced by such approaches, however, as the number of ions detected is not increased relative to the concentration in analyte in the sample.

4.4.6 Resolution and sensitivity

One of the most significant differences between mass analysers is their relative resolving power. As discussed in Chapter 3, quadrupoles have relatively low resolving powers when compared with FTICR and orbitrap mass analysers. The actual resolution achievable, up to the maximum limits set by the analyser, is not fixed, however, and can be tuned by the user. This has a direct effect on the selectivity and also often on the sensitivity of the analysis. In general, the higher the resolution of the measurement, the higher the selectivity, but the lower the sensitivity of a mass spectrometry experiment. This is because to achieve higher selectivity the 'ion beam' or 'packet of ions' inside the mass spectrometer needs to be focused more precisely in order to reduce peak width and increase resolution. To achieve this more focused ion beam, a higher proportion of the ions are lost inside the instrument, reducing the number of ions ultimately transmitted to the detector. Sector instruments and quadrupoles in particular suffer these effects.

We have seen that there is a trade-off between sensitivity, resolution, and selectivity in any mass spectrometry experiment, and that this can be manipulated by the choice of ion source and mass analyser as well as parameter tuning. Whether the most important consideration is mass accuracy or sensitivity, will therefore ultimately determine the optimal choice of mass spectrometer and tuning conditions for a particular mass spectrometry experiment.

4.5 **Summary**

From the material presented in this chapter you should be familiar with the following concepts:

- Resolving power and resolution, and the differences between them.
- The 10% valley and FWHM definitions of resolution.
- Accurate mass measurement, including how it is achieved and its limitations.
- Sensitivity in mass spectrometry, and how this differs from limit of detection.

You should also be aware of the performance capabilities of each type of analyser and be able to compare their advantages and disadvantages.

4.6 **Exercises**

4.1   The following excerpt is taken from a (real) set of teaching notes on TOF mass spectrometry supplied by an A-level examination board. Comment on its accuracy.

'A reflectron is a type of TOF mass spectrometer that enhances the resolution of the mass spectrometer. This is achieved by using a long time-of-flight tube in which the ions move in one direction, are reflected back, and then focused by electrostatic lenses before reaching the detector.'

4.2   A double focusing magnetic sector mass spectrometer is capable of just separating two equally intense ions having m/z values of 272.1068 and 272.1340 with a 10% valley between them.

(i)   What is the resolving power of the analyser?

(ii) What resolution is achieved in the resulting mass spectrum?

Remember to quote all appropriate qualifying information in your answer.

4.3   An FTICR mass spectrometer produces a spectrum with an ion at m/z 454.2596, and a peak width of 0.004 (FWHM). What is the observed resolution to one significant figure?

4.4   MALDI spectra of the protein ubiquitin, shown in the accompanying Figure, are recorded on a TOF MS operating in (a) linear and (b) reflectron modes. Calculate the resolution obtained in each case (to one significant figure), and comment on the significance of these results for reporting the mass of the protein.

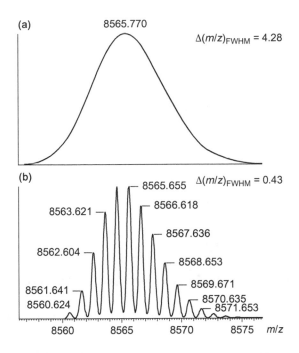

(a)   8565.770

$\Delta(m/z)_{FWHM} = 4.28$

(b)   8565.655 $\Delta(m/z)_{FWHM} = 0.43$

8563.621
8566.618
8562.604
8567.636
8561.641
8568.653
8560.624
8569.671
8570.635
8571.653

8560   8565   8570   8575   m/z

4.5 Metandienone, a performance-enhancing drug, has the molecular formula $C_{20}H_{28}O_2$. Using the data in Table 4.1, calculate the exact mass of:

(i) The neutral molecule.

(ii) The expected molecular ion $M^{+\bullet}$ produced by EI-MS.

(iii) The expected protonated molecular species $[M+H]^+$ produced by ESI-MS.

4.6 A mass spectrometrist is employing accurate mass measurement to determine the presence of pesticide residues on maize crops. The compound of interest is permethrin, molecular formula $C_{21}H_{20}Cl_2O_3$. Using EI-MS, a molecular ion $M^{+\bullet}$ of 390.0782 is measured. Calculate the mass error of this measurement in ppm. What other information from the mass spectrum could be used to corroborate identification of permethrin?

4.7 What is the difference between high resolution and accurate mass measurement in mass spectrometry?

4.8 What two performance properties of a mass spectrometer is sensitivity inversely correlated with?

4.9 Define the terms 'sensitivity' and 'lower limit of detection' with respect to mass spectrometry.

4.10 The negative ion SIM analysis of glutamic acid produced four different total ion counts at four different concentrations in solution. These are recorded in the table below. What was the sensitivity of the method for the analysis of glutamic acid?

Concentration of glutamic acid (ng/mL)	Total ion count
5000	5974 56324
1000	12849 1265
500	58745 632
100	14049 126

4.7 **Further reading**

Chapman, J. R. (1993). *Practical Organic Mass Spectrometry*, 2nd edn. Chichester: Wiley.

Cole, R. B. (ed.) (2010). *Electrospray and MALDI Mass Spectrometry*. New York: Wiley.

Murray, K. K. et al. 'Definitions of terms relating to mass spectrometry (IUPAC Recommendations 2013)', *Pure Appl. Chem.* 85, 1515–609.

5 Tandem mass spectrometry

5.1 Introduction

As well as being able to make accurate m/z measurements from which relative molecular masses can be determined, mass spectrometry is also capable of providing structural information through the generation of fragment ions. Some methods of ionization, such as electron ionization (EI), impart sufficient internal energy into the newly formed molecular ion to promote a significant amount of fragmentation. Other types of ionization, such as electrospray ionization and field ionization, produce ions with very low residual energies and induce little or no fragmentation. In all of these cases, ions can be isolated inside tandem mass spectrometers, according to their m/z, and dissociated to enable examination of the resulting products.

Tandem mass spectrometry, or mass spectrometry/mass spectrometry (MS/MS), is the acquisition and study of the spectra of ions following m/z selection. It is usually combined with a method for inducing ion dissociation by energy transfer (activation). Figure 5.1 shows how an MS/MS spectrum can be generated by isolation and activation of an ion from the corresponding MS spectrum.

The aims of this chapter are to highlight the basic principles of ion fragmentation in a mass spectrometer, examine the methods used to induce dissociation, describe the common tandem mass spectrometer designs, and explain the experiments they make possible.

5.2 Ion dissociation

In addition to an isolation step, MS/MS usually involves ion dissociation, and we will briefly examine the general principles behind this process before looking at the different methods for inducing fragmentation. In order for ions to dissociate they need to possess internal energies above a reaction critical energy (activation energy) E_0, such that chemical bonds can be broken. Ions generated by EI, for example, often possess energies that exceed E_0, and so some fragmentation takes place as a consequence of the ionization process, and peaks corresponding to fragment ions appear in the MS spectrum. Softer ionization methods usually

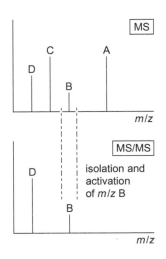

Figure 5.1 The principle of tandem MS. An ion is selected from the MS spectrum and fragmented to produce an MS/MS spectrum.

The ion isolated for study by MS/MS is termed the *precursor ion*, whilst the ions resulting from its reaction or fragmentation are known as *product ions*. In older books you may see the terms 'parent ion' and 'daughter ion' used instead, but these are now out of date.

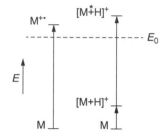

Figure 5.2 Ions generated by EI (left) can often possess energies in excess of the reaction critical energy, and so are able to dissociate in the source. Ions produced by soft ionization methods (right) often require an additional activation step to exceed the energy threshold for dissociation.

require an extra energy transfer step which provides protonated molecules with energies greater than E_0. Figure 5.2 illustrates the concept of reaching the ion-activation energy threshold that leads to fragmentation for each type of ionization process.

When considering the internal energy of ions, and the effect of this energy on dissociation, it is usually more instructive to examine the population of ions as a whole. Now we must use a probabilistic treatment to describe energy distribution, such as that seen in the upper part of Figure 5.3. This shows the probability $P(E)$ of a population of ions possessing some value of internal energy E. At the lower end of the energy distribution we see ions which have insufficient internal energy to dissociate, and are therefore detected as intact molecular ions (EI) or protonated/deprotonated molecules (ESI, MALDI, APCI, etc.). At the higher-energy end, however, we can see ions that possess internal energies $> E_0$, and that can therefore dissociate to the fragment ions F_1^+ and F_2^+. In the case of EI, this fragmentation takes place in the source, whereas with soft ionization methods, such as ESI, it may require additional activation, e.g. in a collision cell (see Section 5.3). For completeness, it should be pointed out that there is a third group of ions, which are termed metastable. They are sufficiently stable to exit the ion source intact, but dissociate before they reach the detector.

Below the probability distribution in Figure 5.3 is a complementary rate plot of log $k(E)$ vs internal energy for dissociation of the molecular ion or protonated/deprotonated molecules to F_1^+ and F_2^+. Notice that, although dissociation to F_1^+

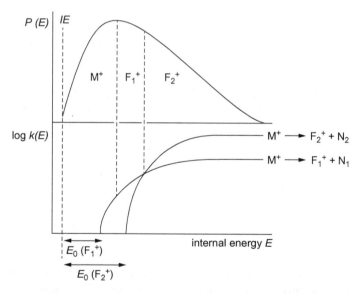

Figure 5.3 A Wahrhaftig diagram showing the relationship between the probability distribution of energies ($P(E)$) for a population of ions and the rate of ion dissociation (log $k(E)$) as a function of energy E. The ion M^+ is a general representation of either the molecular ion $M^{+\bullet}$ or the protonated molecule $[M+H]^+$, depending on the ionization method employed. F_1^+ and F_2^+ represent competing fragment ions, and N_1 and N_2 are the corresponding neutrals.

occurs at a lower value of E_0 than for F_2^+, the latter is the dominant process due to its enhanced rate at higher values of internal energy, E. The result of this is that F_2^+ will be the major fragment ion in the mass spectrum. Plots that link probability distribution and dissociation rates to ion internal energy are known as Wahrhaftig diagrams, after Austin L. Wahrhaftig. They can be used to rationalize fragment ion abundance in a mass spectrum.

For most methods of ion activation, the rate of dissociation is much slower than the rate of energy dissipation throughout the ion. As a consequence, energy is statistically distributed prior to bond cleavage, and the process is said to be thermal, or ergodic. The practical outcome is that weaker bonds tend to fragment to a greater degree than stronger ones. There are, however, a few methods of ion activation where dissociation is extremely fast (see Section 5.3), such that energy is not spread throughout the ion before bond fission (i.e. the process is non-ergodic). This leads to fragmentation modes complementary to those seen by thermal-type processes, and may offer additional analytical information.

5.3 Methods of ion activation and dissociation for MS/MS

Having considered the general principles of ion dissociation we will next examine the most common methods employed to activate ions for fragmentation in MS/MS experiments.

5.3.1 Collision-induced dissociation

Collision-induced dissociation (CID), sometimes known as collisionally activated dissociation (CAD), is the most commonly used method for ion dissociation in MS/MS. Ions are accelerated to high kinetic energy by an electric field, and collided with neutral gas molecules/atoms such as argon, nitrogen, or helium (Figure 5.4). A portion of the precursor ion's kinetic energy is converted into internal energy to induce fragmentation. The total amount of energy transferable is the centre-of-mass energy (E_{com}), which is a function of the ion's kinetic energy (E_{lab}) and the masses of the colliding particles (Equation 5.1).

The total energy available from transfer of translational kinetic E_{lab} to internal energy is the centre-of-mass energy E_{com} given by:

$$E_{com} = \left(\frac{m_N}{m_p + m_N} \right) E_{lab} \qquad (5.1)$$

where m_N is the mass of the neutral gas molecule/atom, and m_p is the mass of the precursor ion. E_{lab} can be calculated from the product of ionic charge z and accelerating voltage used V:

$$E_{lab} = zeV \qquad (5.2)$$

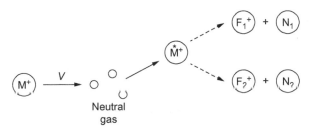

Figure 5.4 Activation and dissociation of an ion M^+ by CID with neutral gas molecules to generate fragment ions F_1^+ and F_2^+, and associated neutrals (N_1 and N_2). The illustration shows a generic positively charged ion. In reality this could be a radical cation, radical anion, or positively or negatively charged molecule (see Chapter 2 for types of ions formed in mass spectrometry).

CID is usually categorized into low-energy collisions ($E_{lab} < 100$ eV), and high-energy collisions ($E_{lab} = 100$ eV–1 keV). The former are usually performed in quadrupole, ion trap, or FTICR-based tandem instruments, whereas the latter are carried out in sector and TOF/TOF spectrometers (see Section 5.4). Low-energy CID usually involves multiple collisions (at higher gas pressures), whereas high-energy CID is typified by a single collision (at lower gas pressures). Dissociation processes that possess large E_0 values are often only accessible in high-energy CID, which can provide information complementary to that available from low-energy collisions.

CID is used in MS/MS of singly-charged small molecules, and multiply-charged biomolecules such as peptides (see Chapters 6 and 8 for applications).

5.3.2 Surface-induced dissociation

This method of ion dissociation resembles CID in that ions are accelerated to high kinetic energy, but an inert solid surface is used as the target instead of neutral gas molecules (Figure 5.5). An obvious difference between SID and CID is the mass of the neutral collision partner. With SID, E_{com} is no longer limited by the mass of the gas molecules, and $E_{com} \approx E_{lab}$. SID is characterized by a single collision producing excited-state precursor ions with a relatively narrow distribution of internal energies. SID offers greater control over energy transfer than CID, since all ions strike the surface at essentially the same angle.

As with CID, SID has been used for MS/MS experiments on both small molecules and biomolecules. Currently, SID is finding application in the study of native multiprotein complexes, where its properties of single collision at well-defined energy provide charge-partition and dissociation modes complementary to those seen by low-energy CID.

5.3.3 Infrared multiphoton dissociation

Infrared multiphoton dissociation (IRMPD) is the most common photodissociation method used for ion activation and dissociation. Precursor ions are trapped and exposed to IR radiation, most commonly from a CO_2 laser operating at 10.6 µm. The stepwise absorption of multiple photons results in activation of vibrational modes within the ion. Rapid redistribution of energy occurs prior to bond cleavage (Figure 5.6) and, as a consequence, thermal-type fragmentation is seen.

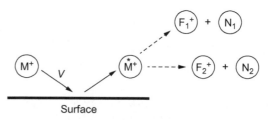

Figure 5.5 Activation and dissociation of an ion M^+ by SID to generate fragment ions F_1^+ and F_2^+, and associated neutrals (N_1 and N_2).

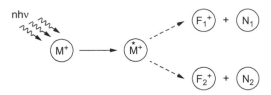

Figure 5.6 Activation and dissociation of an ion M^+ by IRMPD to generate fragment ions F_1^+ and F_2^+, and associated neutrals (N_1 and N_2).

Efficient IRMPD requires precursor ion trapping, and for this reason it is usually performed in FTICR or ion-trap instruments. In these cases, irradiation takes place directly within the mass analyser, via an optical window. IRMPD (and photodissociation in general) is possible in beam-type instruments, where RF-only multipole (usually hexapole) ion guides, with DC trapping lenses, can be used to confine and focus ions during an irradiation step. Depending on the size of the ion cloud, IRMPD typically requires irradiation times of 10–100 ms.

The advantages of IRMPD are the delivery of well-defined energy (0.117 eV per photon at 10.6 μm) and efficient fragmentation. Disadvantages include the need for a vibrational mode capable of absorbing the IR radiation (not normally a problem at 10.6 μm), and the presence of a trapping device in the instrument.

IRMPD can be used to induce MS/MS fragmentation for small and large molecules in a manner similar to low-energy CID.

5.3.4 **UV photodissociation**

Although IRMPD is the most commonly employed laser-based method for ion activation, there has been a recent resurgence in UV photodissociation (UVPD). By using the UV range of the EM spectrum, electronic states of the precursor ion are excited in the first instance. This may lead to direct dissociation of the electronically excited state, or indirect dissociation via dissipation into vibrational modes within the ion. A commonly used wavelength is 193 nm, generated by an ArF excimer laser. At this high frequency each photon delivers 6.4 eV of energy, meaning that single photon absorption is sufficient to induce bond fission in even large protein ions. Indeed, rather like high-energy CID, it allows access to fragmentation pathways that are not seen in low-energy CID or IRMPD.

Like IRMPD, UVPD usually requires some form of ion trapping for efficient operation, but the need for only single-photon absorption reduces irradiation times. Like the other MS/MS ion activated methods discussed so far, UVPD can be applied to both small molecules and large biomolecules. The presence of an aromatic chromophore often enhances fragmentation efficiency, but is not essential, as most organic molecules exhibit some absorption at 193 nm. UVPD is particularly attractive in the analysis of peptides and proteins, where its fast, high-energy activation produces a rich array of fragments and makes it suitable for the analysis of protein modifications (see Chapter 8).

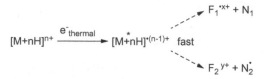

Figure 5.7 Activation and dissociation of an ion $[M+nH]^{n+}$ by ECD to generate fragment ions $F_1^{\bullet x+}$ and F_2^{y+}, and associated neutrals (N_1 and N_2).

5.3.5 **Electron-capture dissociation**

In electron-capture dissociation (ECD), multiply-charged cations interact with low-energy electrons. The ions capture these electrons, resulting in charge reduction and an increase in energy. Removal of an unpaired electron from a multiply-charged odd-electron cation $[M+nH]^{\bullet n+}$ requires 5–7 eV, hence, if an electron is captured by an even-electron species $[M+nH]^{n+}$ to give $[M+nH]^{\bullet (n-1)+}$, 5–7 eV of energy is deposited into the ion, which can induce dissociation (Figure 5.7).

The ECD fragmentation process is fast, such that dissociation occurs before energy is distributed throughout the ion. Under these non-ergodic conditions, strong bonds may be broken in the presence of weak ones. Rather like UVPD, this provides complementary information compared to CID and IRMPD.

As has been mentioned already, ECD requires multiply-charged cations, and for that reason its principal application is in the MS/MS fragmentation of peptide and protein ions generated by ESI. The need to trap both cations and electrons in the same region means that ECD is restricted to use in FTICR mass spectrometers (ion traps are not well suited to ECD). Electrons are produced from a glowing tungsten filament and introduced directly into the rear of the ICR cell (see Chapter 3 for details of the FTICR cell) by a small offset voltage.

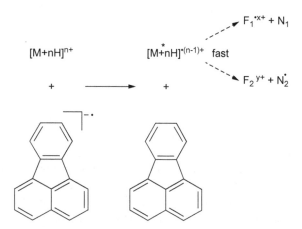

Figure 5.8 Activation and dissociation of an ion $[M+nH]^{n+}$ by ETD using the fluoranthene radical anion to generate fragment ions $F_1^{\bullet x+}$ and F_2^{y+}, and associated neutrals (N_1 and N_2).

5.3.6 **Electron-transfer dissociation**

This method of ion activation and dissociation is similar to ECD in that it delivers electrons to multiply-charged cations to induce fragmentation. The difference between ETD and ECD is that, in ETD, an electron-transfer reagent is used, rather than electrons themselves. The advantage is that it can be performed in ion traps and other trapping devices that are not suitable for ECD. Figure 5.8 illustrates the ETD process with fluoranthene as the transfer reagent. Multiply-charged cations (usually peptides produced by ESI) are allowed to react with fluoranthene radical anions, which are produced by a dedicated ETD chemical ionization (CI) source. Upon transfer of an electron, the charge-reduced, activated radical cation fragments by mechanisms similar to those in ECD.

ETD can be performed in beam instruments, as well as ion traps, providing there is a region where ions can be stored for approximately 100 ms to enable the transfer reaction. Multipole ion guides, or similar, with trapping DC voltages are often used (see Section 5.4). Other reagents used as electron donors in ETD include anthracene and nitrosobenzene.

A summary of the methods of activation and dissociation described in this section is shown in Table 5.1.

Table 5.1 Summary of different ion activation and dissociation techniques and their common areas of application.

Fragmentation approach	Acronym	Energy	Common analyser combinations	Applications	Usage
Collision-induced dissociation	CID/CAD	High & Low	Q-TOF, Q-Orbitrap, Q-FTICR, QqQ, IT, Q-IT	Proteomics, peptide, and small molecule analysis	High
Surface-induced dissociation	SID	High & Low	Q-TOF, Q-FTICR, QqQ, IT, Q-IT	Native multiprotein complexes and small molecules	Low
Infrared multiphoton dissociation	IRMPD	Low	FTICR, IT	Small and large molecules	Medium
UV photodissociation	UVPD	Low	FTICR, IT	Peptides and proteins	Low
Electron-capture dissociation	ECD	Low	FTICR	Peptides, proteins and synthetic polymers	Medium
Electron-transfer dissociation	ETD	Low	IT, IT-Orbitrap, Q-TOF	Proteins and oligomers, phosphoproteomics	Medium

5.4 MS/MS instruments

After considering the different methods of ion activation and dissociation, we will now examine the main types of tandem mass spectrometers used for MS/MS measurement.

5.4.1 Tandem quadrupole analysers

The basic operating principles of a quadrupole mass analyser were described in Chapter 3. Quadrupoles can be combined in series to produce a tandem quadrupole instrument. The most well-known arrangement is the triple-quadrupole, where Q1 and Q3 are used as mass analysers, and q2 (it is convention to designate a non-mass separating quadrupole with lower-case q) is operated in a collision cell in RF-only mode (Figure 5.9). As explained in Chapter 3, if RF-only voltages are applied to a quadrupole, without DC, it focuses and transmits ions of all m/z, and is useful as an ion guide. By placing q2 inside a gas cell it is possible to introduce a collision gas at 10^{-2}–10^{-3} mbar in order to carry out low-energy CID experiments (see Section 5.3). The RF-only quadrupole serves to focus ions after they have undergone collision and greatly increase transmission efficiency.

Most modern tandem quadrupole instruments are not actually triple quadrupoles (QqQ), although they are often colloquially referred to as such. It is now more common to use a hexapole collision cell ion guide instead of q2. This QhQ arrangement affords improved ion transmission and performance. Other ion guide designs, such as stacked rings, have also been employed.

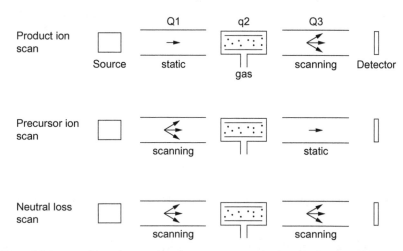

Figure 5.9 Layout of a tandem quadrupole mass spectrometer showing the two resolving quadrupoles Q1 and Q3, and the RF-only q2 in the collision cell. By a combination of static and scanning modes, product, precursor, and neutral-loss MS/MS scans are possible.

MS/MS can be carried out in a tandem quadrupole in one of three common ways: the *product* ion scan, the *precursor* ion scan, and the *neutral loss* scan (Figure 5.9). In *product ion scan mode*, Q1 is set to transmit a particular precursor ion m/z, and Q3 is scanned for the product ions resulting from CID of that precursor in the collision cell. This way the fragment ions originating for the chosen precursor can be studied. In *precursor ion scan mode*, Q3 is set to transmit a particular product ion m/z and Q1 is scanned for precursors that give rise to the product during CID in the collision cell. Precursor ion scanning allows all the precursors of a particular product to be identified. In *neutral ion scanning*, both Q1 and Q3 are scanned with an m/z offset corresponding to a particular neutral loss (e.g. 44 for CO_2). This experiment allows all ions which fragment with that neutral loss to be identified.

Tandem quadrupole mass spectrometers are powerful and flexible instruments, providing high selectivity and sensitivity. They can be coupled to a full range of ion sources, although their limited mass range (approximately m/z 4000) makes their use with MALDI sources impractical. ESI sources are by far the most commonly utilized. Tandem quadrupoles are often employed in combination with HPLC for quantifying trace impurities and metabolites (see Chapters 7 and 8).

5.4.2 **Quadrupole-time-of-flight analysers**

Q-TOFs can be thought of as variants of tandem quadrupoles where Q3 is replaced by a time-of-flight (TOF) analyser (almost always a reflectron TOF). As described in Chapter 3, TOF analysers have a number of advantages over quadrupoles, especially in m/z range and resolving power. The layout of a Q-TOF tandem MS is shown in Figure 5.10. As with tandem quadrupoles, the main form of MS/MS activation is low-energy CID, although commercial instruments are available with ETD capability in the collision cell. Some Q-TOFs have been specially modified for SID or UVPD.

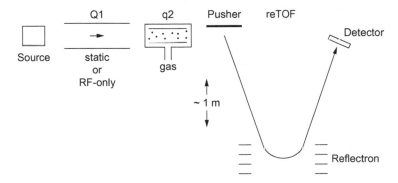

Figure 5.10 Layout of a quadrupole-time-of-flight (Q-TOF) mass spectrometer showing the resolving quadrupole Q1, the RF-only q2 in the collision cell, and the reflectron TOF. In MS mode Q1 is switched to RF-only, and spectra are recorded using the TOF analyser. In MS/MS mode Q1 is set to transmit the precursor for fragmentation in the collision cell and analysis of the product ions by the TOF.

Like tandem quadrupoles, Q-TOFs are capable of product ion scanning, but since the TOF cannot be set to transmit a single *m/z*, precursor ion scanning is not possible. Moreover, since the TOF records spectra on the μs timescale, it is not feasible to link it to Q1 for neutral loss scanning. When Q-TOFs are operated in the non-tandem mode, the quadrupole is set to RF-only, and the TOF acts as the sole mass spectrometer.

The increased *m/z* range and resolving power afforded by *re*TOFs makes the Q-TOF especially suited to biomolecule analysis, where larger ions can be encountered. If ESI is used, then multiply-charged species are produced, with isotope spacing < 1, on the *m/z* scale (see Section 4.2), and high-resolution measurement is required to record monoisotopic mass. If the Q-TOF is equipped with a MALDI source, then the relatively high *m/z* range of the *re*TOF (approximately 20,000) is essential for detecting large singly-charged ions (although the reflectron does place a lower *m/z* limit on the analyser than would be possible with a linear TOF; see Chapter 3).

5.4.3 **TOF/TOF analysers**

This type of tandem instrument is used almost exclusively with MALDI sources, and there are two common types. The first is essentially two TOF analysers in sequence with a collision cell placed between them (Figure 5.11a). In MS/MS mode the first TOF is used to select precursor ions of particular *m/z* based on their flight time taken to reach the collision cell. An ion lens is placed just before the collision cell, and is opened for a narrow time window to allow the selected *m/z* to pass through. Following high-energy CID, the product ions are analysed using the second TOF, usually operated in reflectron mode.

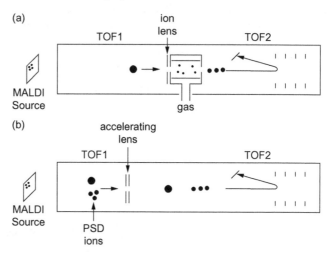

Figure 5.11 Layout of the two common types of TOF-TOF mass spectrometer. In (a), precursor ions are separated in TOF1 and time-selected before being fragmented in the collision cell and the products analysed in TOF2. In (b), ions undergoing post-source decay (PSD) are selected together with their products from TOF1 before reacceleration by an accelerating lens and analysis in TOF2.

The second TOF/TOF design utilizes post-source decay (PSD) of precursor ions to produce an MS/MS spectrum (Figure 5.11b). Ions that are accelerated out of the MALDI source intact, but fragment before they reach the detector, have the same velocity as the precursor and hence the same flight time. If, however, they are subjected to a second round of acceleration, after fragmentation, they will possess different velocities, as they have different masses from the original precursor (see equations 3.8 and 3.9 for the relationships between velocity, time of flight, and m/z). This means that the fragment ions will be detected at a time of flight that corresponds to their own m/z, rather than that of their precursor. Ion selection is achieved by placing the second accelerating lens at a distance beyond the source, and setting it to pulse at a time corresponding to the arrival of the precursor ion and associated post-source decay products. Thus, rather unusually in this type of TOF/TOF, ion dissociation occurs before the selection step.

5.4.4 Ion-trap analysers

As was briefly mentioned, when discussing the principles of the ion-trap analyser (in Chapter 3), it is possible to perform MS/MS in a single ion trap without the need for a second analyser. This is because ion traps are able to isolate ions of a chosen m/z value by removing all other species from the trap. Thus, MS/MS isolation is performed in time rather than space. In simple terms, ion isolation is achieved by applying an appropriate RF combination that ejects ions above and below the desired precursor m/z window. Once isolated, the precursor ion can be activated by a range of methods, including CID, ETD, IRMPD, and UVPD.

Ion-trap CID is most commonly achieved by a technique known as *resonance frequency excitation*. In the case of three-dimensional ion traps, this is related to resonance ejection described in Chapter 3. For linear ion traps, a small AC voltage at resonant frequency to the ion motion is applied directly to the trap rods. For both three-dimensional and linear traps, ion excitation results in collisions with the helium buffer gas at elevated energies. Once the internal energy of the precursor ion exceeds the reaction critical energy, E_0, dissociation occurs. The stepwise process of ion-trap CID MS/MS is shown in Figure 5.12.

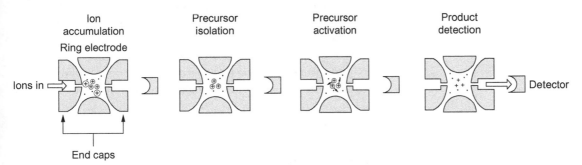

Figure 5.12 Stepwise MS/MS in a three-dimensional ion trap. Following accumulation of all ions, the desired precursor is selected by application of appropriate RF amplitudes. Activation is achieved by application of a resonant frequency that results in higher-energy collisions between the ions and buffer gas. The resulting product ions can then be detected by m/z-dependent ejection onto the detector.

CID in ion traps, although of the low-energy type, differs from that seen in tandem quadrupoles and Q-TOFs, as resonance frequency excitation is *m/z*-specific. This means that once the precursor has dissociated, the fragment ions fall outside the activation window and rapidly 'cool'. In beam-type instruments, fragment ions may still undergo further, relatively high-energy, collisions and dissociation within the collision cell.

In addition to CID, ion traps are also well suited to ETD and laser-based dissociation techniques. ETD reagent radical anions can be injected into the trap to react efficiently with previously selected, multiply-charged, cation precursor ions. Laser irradiation of ions can be performed in three-dimensional traps via a window in one of the end-cap electrodes. In linear ion traps this is achieved by irradiation along the *z*-axis (see Figure 3.6). This makes both IRMPD and UVPD possible in ion traps.

5.4.5 FTICR analysers

Rather like ion traps, FTICR instruments do not necessarily require an additional analyser to isolate precursor ions. Following initial trapping of the total ion population in the ICR cell, an intense broadband radio-frequency pulse may be applied. The pulse has a 'gap' at the cyclotron frequency corresponding to the precursor *m/z* of interest, meaning that all ions except those of the desired *m/z* are ejected from the cell. Following precursor isolation, a number of ion activation methods are possible in the cell, including CID, ECD, IRMPD, and UVPD.

CID is usually achieved in the ICR cell by a technique known as sustained off-resonance irradiation-collisionally activated dissociation (SORI-CAD). A relatively long RF-pulse, very close to the resonance frequency of the precursor ion, is applied whilst a burst of collision gas (usually argon) is injected into the ICR cell. This causes an increase in the kinetic energy of the precursor ion and results in fragmentation through collisional processes. This was one of the earliest types of MS/MS to be used in FTICR instruments, but it has the disadvantage of introducing gas into the high vacuum region of the analyser, which can compromise resolving power. With the introduction of hybrid FTICR mass spectrometers, CID is now more commonly performed outside the ICR cell (see Section 5.4.6).

MS/MS ion activation methods that are still commonly performed in the ICR cell include ECD, IRMPD, and UVPD. The trapped ion cloud is ideal for sustained interactions with thermal electrons or for laser irradiation. For ECD, electrons are released from a heated cathode and introduced into the rear of the cell by application of an offset voltage. To facilitate laser irradiation, a window is usually incorporated into the back plate of the instrument's flight tube so that ions can be directly irradiated in the cell.

5.4.6 Other combinations of analysers

In addition to triple quadrupole and Q-TOF tandem mass spectrometers, other hybrid instruments employing a 'front-end' quadrupole include Q-FTICR and Q-orbitrap variants. These designs also incorporate a post-quadrupole collision cell to allow fragmentation of selected precursor ions, but offer the very high resolving power and mass accuracy for product ion analysis afforded by FTMS.

The linear ion trap provides a powerful 'front-end' analyser for hybrid mass spectrometers, and has been used in combination with TOF, FTICR, and orbitrap instruments. The trap is capable of precursor isolation and activation for CID MS/MS. Product ions can either be ejected from the sides of the trap onto a detector (i.e. acting as a stand-alone ion trap), or passed through the rear of the trap to the TOF, FTICR, or orbitrap analyser for higher-resolution analysis. The ion trap also has the advantage of being able to accumulate defined ion populations, which means that overloading of the down-stream FTICR cell or orbitrap, with associated space-charge effects, can be avoided.

5.5 MS/MS experiments

We saw in Section 5.4 that tandem analyser configurations can be used in a variety of ways to provide information about analytes present in a sample. For example, tandem quadrupole analysers can be used for 'product ion scanning', 'precursor ion scanning', and 'neutral loss scanning'. There are several additional tandem mass spectrometry experiments that are commonly used for analyte identification, enhancing selectivity and quantitation. The following section provides selected examples.

5.5.1 MS^n

If we consider the way in which tandem mass analysers function there are fundamentally two approaches: the *tandem in space* configurations which include ion beam analysers, such as triple quadrupoles, and *tandem in time* configurations found in ion traps and FT-ICR systems. One of the advantages of the 'trapping' analysers is that they can be used for successive fragmentation of product ions via an experiment referred to as MS^n (Figures 5.13a and 5.13b). Being able to select a specific precursor ion and subject it to several generations

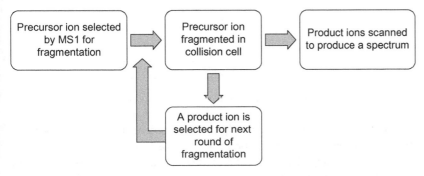

Figure 5.13a The successive fragmentation of product ions from a single precursor selection using MS^n.

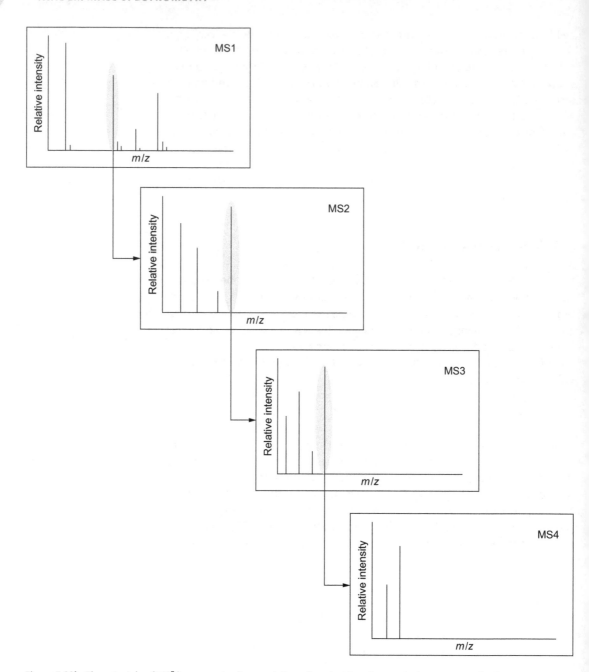

Figure 5.13b The principle of MS^n by successive fragmentation of product ions from a single precursor selection.

of product ion fragmentation clearly provides potential for generating more structural information, when compared to what is available from a single product ion spectrum. This tandem MS tool provided by trapping analysers is especially useful for structural elucidation of unknown compounds and the differentiation of structural isomers.

Although MS^n has the potential to provide unprecedented information to help elucidate the structure of unknown compounds, tandem mass spectra can also be invaluable in helping identify previously characterized compounds. The high reproducibility of EI fragmentation is not shared by MS/MS spectra derived from soft ionization techniques, hence universal libraries of product ion spectra (akin to those used in EI) are not widely used. However, when combined with additional measurement criteria, such as accurate mass measurements, MS/MS spectra (MS^2 spectra) can add specificity to the interpretation of compound identity.

5.5.2 Data-dependent experiments (DDA) and data-independent experiments (DIA)

DDA MS/MS measurements provide product ion spectra for multiple analytes by automatically switching from MS to MS/MS mode. Real-time data is used to determine which precursor ions are fragmented and for how long, which makes it particularly well suited to experiments using separation techniques combined with mass spectrometry. Within the MS method, criteria are set for selecting precursors and can be modified by the user. These can include factors such as setting an ion abundance threshold and the number of precursor ions that can be selected per MS scan (this may be typically 1–20, but the acquisition rate of the analyser is a key determinant and can become a limiting factor as the number of precursors increase). A predefined length of time for the accumulation of product ion scans can be selected, or alternatively a target ion abundance set. Clearly, the more precursor ions that are selected per MS scan, and the higher the product ion abundance target, the more time the instrument will spend in MS/MS mode and the less time will be spent scanning for new precursors. DDA methods can be used to produce large numbers of product ion spectra from complex samples, but MS/MS coverage may not be comprehensive, as only a limited number of compounds can be fragmented per MS scan, and careful setup is required to ensure that ions of interest are not excluded. Figure 5.14 provides a schematic of the DDA processing showing an MS1 scan triggering five MS2 scans, with this process running sequentially over time.

Data-independent analysis (DIA) shares some similarities with DDA. DIA experiments also alternate between MS and MS/MS data acquisition modes, but the transition from one to the other is not determined by the compounds in the mass spectrum. In contrast, DIA alternates the scan function from MS to MS/MS based on a fixed period of time for each scan type, and the MS/MS process does not focus on single or consecutive precursor ions for fragmentation. Instead, the first analyser transmits all precursors eluting at that point in time across the full m/z range, and all ions are fragmented simultaneously under fixed conditions in the fragmentation cell. This means that the MS/MS spectra contain product ions from all possible precursors. By alternating between MS and MS/MS acquisitions in this

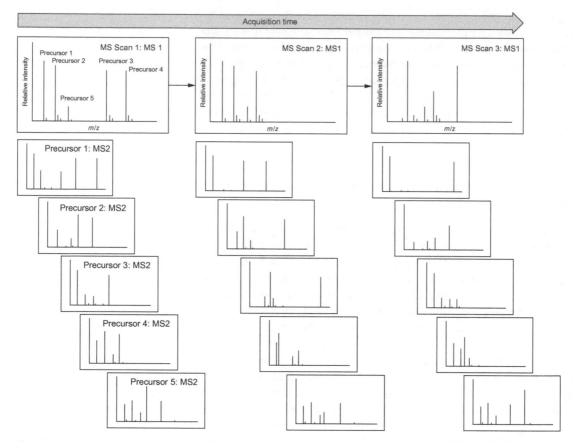

Figure 5.14 Using data-directed analysis (DDA), a single scan on the first analyser (MS1) selects a defined number of ions for sequential fragmentation (five in the example). MS2 (MS/MS) spectra are then acquired consecutively from the product ions of each precursor. When complete, the next MS scan takes place and the process is repeated throughout the analysis of the sample. DDA is of benefit when the sample composition changes over time, for example when chromatography is coupled directly with mass spectrometry.

way, potentially more product ions are generated from a greater number of precursors compared to the DDA method. Software deconvolution enables MS and MS/MS spectra to be matched in time. The number of product ions formed from each precursor is not monitored, however, and hence higher-abundance precursors will tend to be better represented in the product ion spectrum for any particular MS/MS scan.

Both DDA and DIA experiments can be particularly useful for profiling complex environmental and biological samples. For example, in metabolomics (see Chapter 8) tandem mass spectra from DDA and DIA experiments are used to add confidence to compound identifications and to help identify unknown metabolites. Similarly, in proteomic profiling (see Chapter 8) the product ion MS/MS spectrum of a sequence of peptides in a sample, derived from the tryptic digest of a protein, can use DDA or DIA to confirm the amino acid sequence of each peptide and determine a protein's identity in a single experiment.

5.5.3 **Selective reaction monitoring (SRM)**

So far we have seen how tandem mass spectrometry can be used to gain structural information and to help in compound identification. Next we shall discuss how it can be used to improve selectivity, reduce detection limits, and enhance quantitative analysis.

An SRM experiment can be performed on certain tandem mass spectrometers, most commonly triple quadrupole systems. The first analyser is set to transmit a specific *m/z* value which corresponds to the ions of a particular analyte of interest. These selected precursor ions are then fragmented in a collision cell, and all the products are transferred to a second analyser which is set to transmit only a specific *m/z* value from the product ion spectrum. Figure 5.15 provides a schematic of the SRM process. Precursor and product *m/z* values are chosen prior to the experiment, and may be the base peak or most diagnostic peak of significant abundance in the product ion spectrum. This controlled fragmentation reaction is referred to as a *transition* and has the following useful characteristics: only a single product ion from the precursor is transmitted to the detector, so the experiment is highly selective, and is more selective than the SIM experiment (see Chapter 4) because SIM does not distinguish between different compounds that have the same *m/z* value within the SIM window, but SRM does.

1. Because only the product *m/z* value is transmitted to the detector, the chemical background is as low as it can be within the confines of the performance of the analyser. This increases the signal-to-noise and lowers the detection limit. It is worth noting that the total number of product ions generated is almost always less than the number of precursor ions (due to < 100% fragmentation efficiency). Sensitivity is therefore not technically increased, but the background is generally reduced to a greater extent than the MS/MS transmission efficiency losses, and so the detection limit can be lowered.

2. The *m/z* scale is not scanned during the SRM experiment, and so the product ion can be monitored continuously and no data are lost whilst scanning an

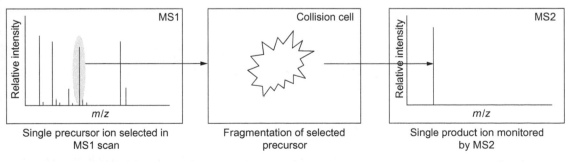

Figure 5.15 In single-reaction monitoring (SRM) a precursor ion is selected for fragmentation by the first analyser (MS1). The second analyser (MS2) is then set to monitor a single *transition* product ion which is diagnostic of the precursor. This enhances selectivity and is useful for quantitation.

m/z range (i.e. a duty cycle of 100%). This has significant benefits when monitoring low-abundance compounds using LC-MS or GC-MS configurations (see Chapter 7).

SRM experiments are not limited to a single reaction, and most instrument control software will enable a number of reactions to be monitored throughout an acquisition. Note that this is often referred to as *multiple reaction monitoring* (MRM), although this term was in fact introduced for commercial reasons and is perhaps not entirely necessary given than it refers simply to an experiment in which a number of SRM experiments take place consecutively during an acquisition. The multiple SRM experiment can be useful for LC-MS or GC-MS where a number of different compounds of interest may elute across a chromatographic run.

As well as providing what is often a significant enhancement in limit of detection, SRM experiments can be very useful for quantitative analyses. When using full scan or SIM acquisitions for quantitative experiments, co-eluting matrix components at the same *m/z* value as the analyte will falsely contribute to the peak area of the 'analyte signal'. Even if these matrix components co-elute and have the same *m/z* as the analyte they are unlikely to have the same transition product *m/z* value, so the SRM experiment can be used to remove their contribution to the analyte response. The longer time spent monitoring the transition also means that often more data points per chromatographic peak can be recorded (see Chapter 7). The lower the overall duty cycle of the instrumentation, the more an SRM-approach can help enhance the reproducibility of peak area measurements. One of the reasons why triple quadrupoles in particular are favoured for quantitative experiments is their very fast switching, which can be combined very effectively with multichannel SRM experiments.

5.6 Summary

From the material presented in this chapter you should be familiar with the following:

- The meaning and significance of tandem mass spectrometry (MS/MS).
- The basic principles of ion dissociation.
- Methods of ion activation for MS/MS.
- The common tandem mass spectrometer designs.
- The principles and analytical applications of SRM and MRM measurement.

5.7 Exercises

5.1   What is tandem mass spectrometry, and what does an MS/MS spectrum show?

5.2   Why do ions produced by ESI often require additional activation in order to dissociate?

5.3   What factors affect the presence and abundance of fragment ions in an MS, or MS/MS, spectrum?

5.4   Calculate the centre-of-mass energy (E_{com}) in eV for a doubly-charged ion of mass 1150.5 Da activated by collision with an argon atom (mass 39.95 Da) following acceleration using a collision voltage of 35.0 V.

5.5   What would be the approximate E_{com} (in eV) of the ion in Question 5.4 if SID were used instead of CID? Comment on the difference in energies.

5.6   The reaction critical energy, E_0, for dissociation of an ion is 3.50 eV. Determine the minimum number of photons that must be absorbed to induce fragmentation by:

   (i)   IRMPD using a wavelength of 10.6 μm,

   (ii) UVPD using a wavelength of 193 nm.

5.7   List two advantages of a Q-TOF mass spectrometer over a tandem quadrupole instrument. Does a tandem quadrupole have any capabilities that are not typically exhibited by a Q-TOF?

5.8   What is the principal difference between CID in tandem quadrupole instruments and that in ion traps?

5.9   Why is it often beneficial to conduct CID external to an FTICR cell, rather than use SORI-CAD?

5.10 Which types of mass analyser can be used for MS^n experiments?

5.11 Explain how an SRM experiment can lead to greater selectivity and contribute to more accurate quantification of analytes.

5.12 Discuss some of the advantages and disadvantages of using DDA compared to DIA in an LC/MS experiment to identify and quantify metabolites in blood plasma.

5.8 Further reading

Boyd, R. K., Basic, C., and Bethem, R. A. (2008). *Trace Quantitative Analysis by Mass Spectrometry*. Chichester: Wiley.

Cole, R. B. (ed.) (2010). *Electrospray and MALDI Mass Spectrometry*. New York: Wiley.

Hoffmann, E. de and Stroobant, V. (2007). *Mass Spectrometry: Principles and Applications*, 3rd edn. Chichester: Wiley.

Watson, J. T. and Sparkman, O. D. (2007). *Introduction to Mass Spectrometry*, 4th edn. Chichester: Wiley.

Interpretation of mass spectra

6

6.1 Introduction

As we have seen in Chapter 5, molecular ions ($M^{+\cdot}$), or protonated/deprotonated molecules ($[M+H]^+$/$[M–H]^-$), can undergo dissociation if they possess sufficient internal energy. The resulting fragmentation pattern seen in the mass spectrum (or MS/MS spectrum) can be used to provide information about the structure of the molecule. In this chapter we will examine how fragmentation data can be interpreted. The main concepts will be highlighted and examples provided.

The first part of this chapter will outline the fragmentation of small organic molecules using EI-MS. Following an introduction of the basic types of fragmentation process, the diagnostic ions and neutral losses seen for different organic structures and functional groups will be presented. The second part of the chapter will summarize the dissociation of protonated and deprotonated molecules ($[M+H]^+$/$[M–H]^-$), such as those formed by ESI, MALDI, or CI. Here, unlike EI, additional activation is often required to induce fragmentation (as described in Chapter 5). We will discuss the fragmentation of small molecules as well as peptides and other large biomolecules.

6.2 Interpretation of EI mass spectra

6.2.1 Location of the charge/radical

One of the first considerations in interpreting the EI fragmentation behaviour of an ion is where the charge resides. The majority of bond fission is centred around the charge/radical site. Clearly, even within a modest structure containing tens of atoms, there may be numerous potential charge sites. In EI we are concerned with the formation of molecular ions ($M^{+\cdot}$) by removal of an electron, and it is most likely that this will be lost from the highest (energy) occupied molecular orbital (HOMO). Non-bonding electrons are generally highest in energy, so if the molecule contains atoms with lone pairs, such as N, O, S, halogen, etc., it is probable that an electron will be removed from one of these atoms,

The general order of increasing orbital energy in a molecule is sigma (σ) < pi (π) < non-bonding (n).

in preference to bonding electrons, and that the resulting charge and unpaired electron within the molecular ion will reside at this location. This is not an exact rule, and other processes, such as removal of electrons from σ and π orbitals, will almost certainly compete to drive additional modes of fragmentation. Nonetheless, as we shall see, this simple assumption is very useful when interpreting EI mass spectra.

6.2.2 Molecular ions, fragment ions, and neutral losses

As we saw in Chapters 1 and 2, EI mass spectra usually contain a molecular ion and multiple fragment ions derived from the loss of neutral species (radicals and molecules) resulting from a fragmentation cascade. For the purposes of interpretation it would be helpful to establish whether a particular ion is an $OE^{+\bullet}$ or EE^+ species, and also whether a given neutral loss is due to removal of a radical or an even electron fragment. Luckily, there is a general rule that helps us to determine these differences.

The *nitrogen rule* states that organic molecules containing any of the elements C, H, N, O, P, S, Si, and any halogens will have (i) an *odd* monoisotopic molecular mass if they contain an *odd* number of nitrogen atoms (N_1, N_3, N_5, etc.), or (ii) an *even* monoisotopic molecular mass if they contain an *even* number of nitrogen atoms (N_0, N_2, N_4, etc.). For the purposes of this rule the *nominal monoisotopic molecular mass* is used (i.e. the mass rounded to the nearest integer value).

What does this mean for the interpretation of mass spectra? Given that a molecular ion is just a neutral molecule minus an electron, it follows that it will also obey the nitrogen rule. In fact this holds for all $OE^{+\bullet}$ ions in a spectrum. EE^+ ions, in contrast, do not have the formula of a neutral molecule and show the exact opposite relationship between odd/even mass and the number of nitrogen atoms.

Consider the simplest case where there are no nitrogen atoms in the molecule: the molecular ion, and all $OE^{+\bullet}$ ions, will have even m/z values, whereas all the EE^+ ions will have odd m/z values. This is illustrated in Figure 6.1, which shows the EI mass spectrum of heptan-3-one. The two $OE^{+\bullet}$ ions at m/z 114 and 72 are readily identifiable as odd electron ions by their even m/z values, whereas the three ions at (odd) m/z 85, 57, and 29 must be EE^+ ions. We will return to further interpretation of this spectrum later.

The position becomes a little more complicated when the molecule contains one or more nitrogen atoms. For the molecular ion it is relatively straightforward, as this has the same formula as the parent molecule, and, as already stated, the rule always applies. Thus, for a molecule containing an odd number of nitrogen atoms, $M^{+\bullet}$ will always be an odd number, and for a molecule with an even number of nitrogen atoms, $M^{+\bullet}$ will always be an even number. The complication arises for the fragment ions, as they may contain some or none of the original nitrogen atoms from the molecule. An even m/z fragment ion may be an EE^+ ion with an odd number of nitrogen atoms, or an $OE^{+\bullet}$ ion with an even number of nitrogen atoms.

Odd electron ions $OE^{+\bullet}$ are radical cations, possessing both a positive charge and an unpaired electron. Even electron ions EE^+ are cations with paired electrons. In EI the molecular ion ($M^{+\bullet}$) is always an odd electron species, but fragment ions may possess an odd or even number of electrons, depending upon the mechanism of their formation. EE^+ are usually dominant in the spectrum.

Odd electron ions $OE^{+\bullet}$ can dissociate to produce either $OE^{+\bullet}$ fragment ions, by loss of a neutral molecule, or EE^+ fragment ions, by loss of a neutral radical. In contrast, except for a few special cases, EE^+ ions can only lose neutral molecules to give EE^+ fragment ions.

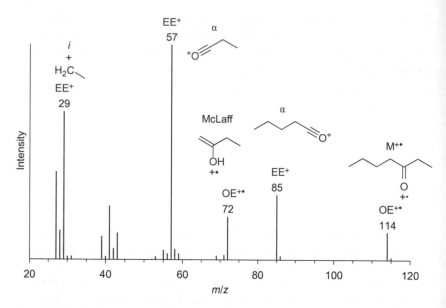

Figure 6.1 The EI mass spectrum of heptan-3-one, a molecule containing no nitrogen atoms. The two OE$^{+\bullet}$ ions can be identified by their even m/z values, and the three EE$^+$ ions by their odd m/z values.

This problem of distinguishing the nitrogen content of fragment ions is partially overcome by the tendency of EE$^+$ ions to outnumber OE$^{+\bullet}$ ions, which can allow initial assumptions to be made. Access to accurate mass measurement, and thereby elemental composition of ions (see Chapter 4), allows unequivocal determination of the number of nitrogen atoms.

6.2.3 Energetic factors and fragment ion abundance

Ion fragmentation during EI is a *unimolecular* process. In other words, the ion undergoing dissociation does not interact with any other ions or neutral molecules during the process, due to the low pressures in the EI ion source. Fragmentation can also be considered an irreversible process for the same reason. As we saw in Chapter 5, an ion must possess an internal energy greater than the reaction critical energy E_0 in order to fragment. Since energy is distributed throughout the ion prior to dissociation, weaker bonds will tend to dissociate in preference to stronger ones. This is not the only consideration in determining fragment ion abundance, however, as primary fragments may still possess sufficient internal energy to dissociate further. Thus, the stabilities of the fragment ion and the neutral lost are important factors. Those ions that are better able to stabilize a positive charge will tend to fragment less readily and be more abundant in the spectrum. Loss of a stable neutral molecule by-product (e.g. CO, HCN, C_2H_4) or a stabilized radical by-product (allylic, benzylic, etc.) will also tend to promote the abundance of a given production.

Another useful guide is *Stevenson's rule*. This states that when an ion dissociates, the charge will tend to remain on the fragment with the lower *ionization energy*.

Despite its name this is not an absolute rule, and there are competing factors, such as the tendency to lose larger radicals preferentially.

6.2.4 Basic fragmentation mechanisms

We will consider four basic types of fragmentation mechanism: σ-bond dissociation, charge-mediated fragmentation, α-cleavage, and rearrangements. These are sufficient to explain formation of most of the fragment ions seen in EI mass spectra.

σ-bond dissociation occurs principally when the electron lost during ionization comes from a single bond (Figure 6.3). Although these electrons tend to be of lower energy, they can be removed during EI, and the mechanism does give rise to a significant amount of fragmentation. It is the major mode of dissociation for alkanes, for example (see Section 6.2.5).

Charge-mediated fragmentation is induced by the charge site and leads to *heterolytic* cleavage of the bond adjacent to the charge site (Figure 6.4). It is also known as *heterolytic fragmentation* or *inductive cleavage*, and is symbolized with the letter *i*. The electrons move to the location of the cation, which causes migration of the charge. It can occur with both odd and even electron ions, and at saturated or unsaturated sites.

α-cleavage is *homolytic* fission of the bond between the atom adjacent to the radical site and the atom two bonds away from the radical site (i.e. between atoms 2 and 3). It is induced by the radical, and involves the movement of single

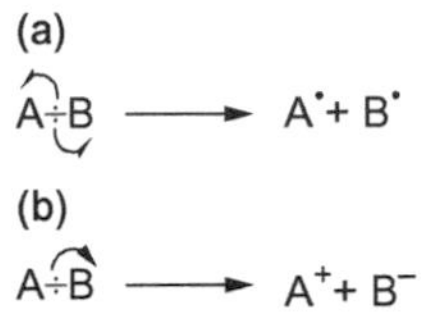

(a)

$$A \dot{\div} B \longrightarrow A^{\bullet} + B^{\bullet}$$

(b)

$$A \dot{\div} B \longrightarrow A^{+} + B^{-}$$

Figure 6.2 Reminder: (a) homolytic or (b) heterolytic bond cleavage of a neutral molecule A-B leads to a pair of radicals, or a cation and anion pair, respectively. Single-headed or double-headed 'curly arrows' are used to show the movement of one or two electrons respectively.

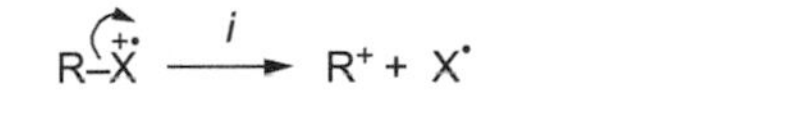

$$RCR_3 \xrightarrow{EI} R\dot{\div}CR_3 \xrightarrow{\sigma} R^{\bullet} + {}^{+}CR_3$$

Figure 6.3 EI ionization and σ-bond dissociation leading to an even electron cation and loss of a radical.

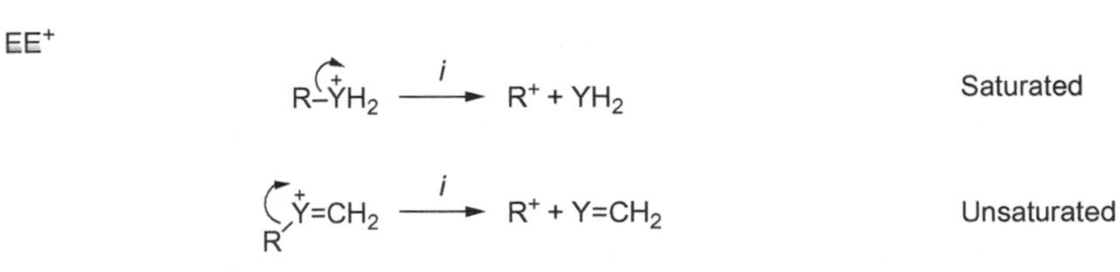

$OE^{+\bullet}$

$$R\widehat{-X}^{+\bullet} \xrightarrow{i} R^{+} + X^{\bullet} \qquad \text{Saturated}$$

$$\xrightarrow{i} R^{+} + \left(\underset{R}{\overset{\bullet\bullet}{}} {-} Y^{\bullet} \longleftrightarrow \underset{R}{} {=} Y \right) \qquad \text{Unsaturated}$$

EE^{+}

$$R\widehat{-Y}H_2^{+} \xrightarrow{i} R^{+} + YH_2 \qquad \text{Saturated}$$

$$\underset{R}{\overset{+}{\widehat{C}}}{=}CH_2 \xrightarrow{i} R^{+} + Y{=}CH_2 \qquad \text{Unsaturated}$$

Figure 6.4 Charge-mediated fragmentation of odd and even electron ions at saturated and unsaturated sites. X = OH, OR, NR$_2$, halogen, etc.; Y = O, NR, etc.

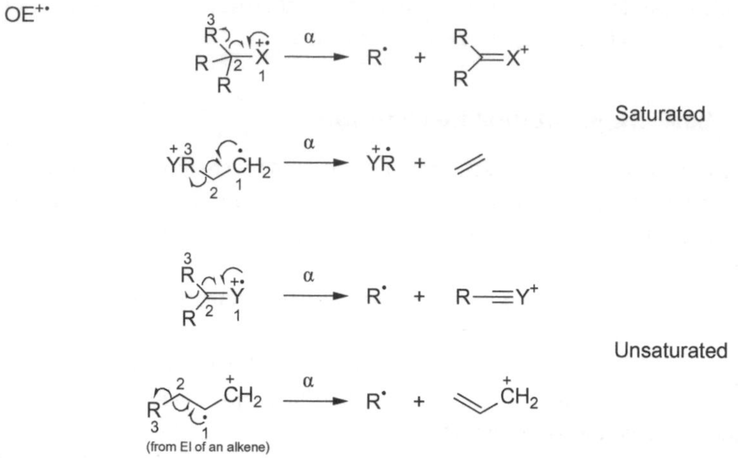

Figure 6.5 α-cleavage at saturated and unsaturated positions leads to cleavage of the bond between the atom adjacent to the radical site and the atom two away from the radical site. X = OH, OR, NR$_2$, halogen, etc.; Y = O, NR, etc.

electrons. α-cleavage is a major fragmentation pathway of OE$^{+\bullet}$ ions, and can occur at saturated and unsaturated positions (Figure 6.5). It is noteworthy that charge-mediated fragmentation and α-cleavage often (but not always) lead to complementary pairs of ions, such that where R$^+$ may be observed in the former mechanism, R$^\bullet$ is lost as a neutral in the latter.

Rearrangements are the final type of fragmentation we will consider. There are numerous ways in which dissociation can occur following migration of atoms or groups within an ion. It is beyond the scope of this book to cover these in detail, and so we will just consider some of the more important ones. Rearrangements can be subdivided into those initiated by the radical site, and those initiated by the charge site. Of the former, the most well-known is the McLafferty rearrangement, named after the eminent mass spectrometrist Fred W. McLafferty. It involves a 1,5-migration of a hydrogen radical (i.e. via a six-membered cyclic transition state) to an unsaturated site, followed by fission of the bond between atoms 3 and 4 (see Figure 6.6). Two cleavage mechanisms are possible: a homolytic, α-cleavage, type where the charge is retained on its original site, and a heterolytic, charge-mediated type where the charge migrates from its original site.

Hydrogen radical shifts to saturated radical sites are also common, and these often include 1,3- and 1,4- as well as 1,5-shifts (although migrations from other positions are also possible). Following the rearrangement step, ion dissociation may occur by α-cleavage, radical displacement (*rd*), or charge-mediated fragmentation. Radical displacement is analogous to α-cleavage, except that the bond undergoing fission is not that between atoms 2 and 3 (see Figure 6.7).

Finally, we consider charge-site initiated rearrangement. The most common type is hydride migration to the charge site with concurrent elimination of a neutral ring or alkene (see Figure 6.8).

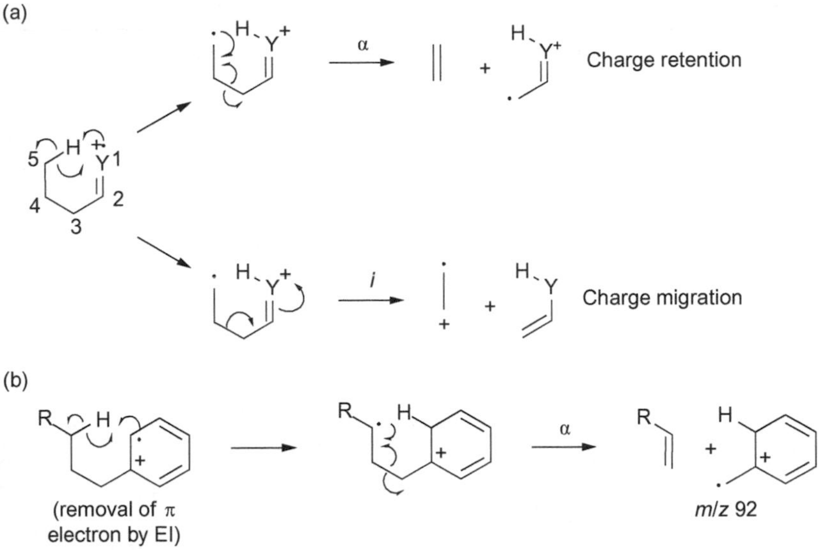

(a)

Charge retention

Charge migration

(b)

(removal of π
electron by EI)

m/z 92

Figure 6.6 The McLafferty rearrangement centred on (a) a heteroatom radical, and (b) a carbon radical. Following a 1,5-hydrogen radical shift, α-cleavage or charge-mediated fragmentation may occur leading to charge retention or charge migration respectively. Y = O, NR, etc.

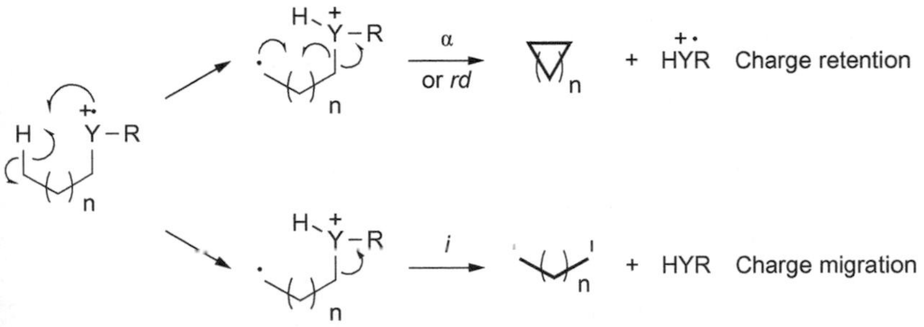

+ HYR   Charge retention

+ HYR   Charge migration

Figure 6.7 Rearrangement by hydrogen radical shift to a saturated site followed by α-cleavage, radical displacement (rd), or charge-mediated fragmentation (i) leading to charge retention or charge migration respectively. Note that n can equal zero. Y = O, NR, etc.

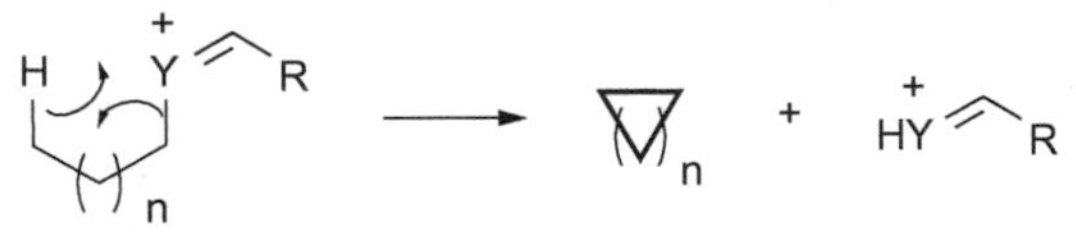

Figure 6.8 Hydride rearrangement to the charge site with concurrent loss of a ring (n > 0) or alkene (n − 0) neutral. The EE⁺ precursor shown may itself have arisen from α-cleavage of an OE⁺, or similar process. Y = O, NR, etc.

6.2.5 Fragmentation of common compound classes

The typical fragmentation behaviour of some organic molecules are described in this section. Additionally, Tables 6.1 and 6.2 provide lists of common fragment ions, and neutral losses, respectively.

Table 6.1 The masses and structures of some common fragment ions (to three decimal places).

m/z[a]	Structure	Interpretation[b]
29.039	$CH_3CH_2^+$	Ethyl group, σ, i
31.018	$H_2C=O^+H$	Primary alcohol, α
43.018	$H_3CC\equiv O^+$	Acetyl group, α
43.055	$[C_3H_7]^+$	Propyl/isopropyl group/alkane, σ, i
45.034	$H_3CHC=O^+H$	2-alcohol, α
48.985/50.982	$H_2C=Cl^+$	Primary chloride, α
55.055	$[C_4H_7]^+$	Butenyl group/alkene, σ, i
57.034	$H_3CH_2CC\equiv O^+$	Propanoyl group, α
57.071	$[C_4H_9]^+$	Butyl group/alkane, σ, i
58.042	+• OH (2-ketone enol structure)	2-ketone, McLafferty
60.021	+• OH, OH (carboxylic acid enol structure)	Carboxylic acid, McLafferty
69.071	$[C_5H_9]^+$	Pentenyl/alkene, σ, i
71.050	$H_3CH_2CH_2CC\equiv O^+$	Butanoyl group, α
71.086	$[C_5H_{11}]^+$	Pentyl/alkane, σ, i
72.058	+• OH (3-ketone enol structure)	3-ketone, McLafferty
74.037	+• OH, OCH_3 (methyl ester enol structure)	Methyl ester, McLafferty
77.039	phenyl cation structure $^+$	Phenyl group, α
78.034	pyridyl cation structure $^+$	Pyridyl group, α
83.086	$[C_6H_{11}]^+$	Hexenyl group/alkene, σ
85.065	$H_3CH_2CH_2CH_2CC\equiv O^+$	Pentanoyl group, α
85.102	$[C_6H_{13}]^+$	Hexyl group/alkane, σ, i
86.073	+• OH (4-ketone enol structure)	4-ketone, McLafferty
88.053	+• OH, OCH_2CH_3 (ethyl ester enol structure)	Ethyl ester, McLafferty

Table 6.1 (Continued)

91.055		Benzyl group, α, σ, i
92.934/94.932	$H_2C=Br^+$	Primary bromide, α
105.034		Benzoyl group, α

^{a} m/z for monoisotopic mass to 3 d.p. ^{b} Including likely fragmentation mechanism.

Table 6.2 The masses and structures of some common neutral losses (to three decimal places).

$\Delta m/z^a$	Structure	Interpretation^b
1.008	H˙	Loss of H from e.g. aldehyde, α
15.024	$H_3C^˙$	Loss of methyl group, α, σ
17.027	NH_3	Amine, H-migration then i
18.011	H_2O	Alcohol, H-migration then i
20.006	HF	Fluoride, H-migration then i
26.016	HC≡CH	Eliminated from alkene or aryl
27.995	CO	Butenyl group/alkene, σ, i
28.031	$H_2C=CH_2$	Eliminated from e.g. alicyclic ring
29.003	$H˙C=O$	Aldehyde
29.039	$H_3CH_2C^˙$	Loss of ethyl group, α, σ
31.018	$H_3CO^˙$	Loss of methoxy group from e.g. methyl ester, α
34.969/36.966	Cl˙	Chloride, i
41.039	$H_2C=CHH_2C^˙$	Loss of propenyl group, α, σ
43.055	$H_3CH_2CH_2C^˙$	Butanoyl group, α
71.086	$[C_5H_{11}]^+$	Pentyl/alkane, σ, i
72.058	–	3-ketone, McLafferty
74.037	–	Methyl ester, McLafferty
77.039	–	Phenyl group, α
78.034	–	Pyridyl group, α
83.086	$[C_6H_{11}]^+$	Hexenyl group/alkene, σ
85.065	$H_3CH_2CH_2CH_2CC≡O^+$	Pentanoyl group, α
85.102	$[C_6H_{13}]^+$	Hexyl group/alkane, σ, i
86.073	–	4-ketone, McLafferty
88.053	–	Ethyl ester, McLafferty
91.055	–	Benzyl group, α, σ, i
92.934/94.932	$H_2C=Br^+$	Primary bromide, α
105.034	–	Benzoyl group, α

^{a} m/z for monoisotopic mass to 3 d.p. ^{b} Including likely fragmentation mechanism.

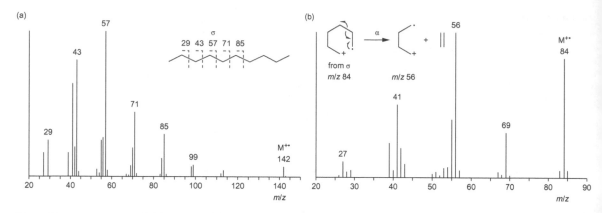

Figure 6.9 EI-mass spectra of (a) decane and (b) cyclohexane.

Alkanes generally give rather weak but detectable molecular ions that cleave by σ-bond dissociation to give a distribution of fragments of the type $[C_nH_{2n+1}]^+$ at m/z 29, 43, 57, 71, 85, etc. (see Figure 6.9a). Branched alkanes exhibit preferential bond cleavage either side of the branching site, due to the increased stability of the product ion and/or neutral produced. This is an analytically useful feature, as it can allow the branching site to be deduced.

Cyclic alkanes tend to give stronger molecular ions than their acyclic counterparts, and the loss of ethene from the molecular ion is a common fragmentation (σ followed by α; see Figure 6.9b for the example of cyclohexane). Larger rings often show multiple losses of ethene.

Alkenes tend to show a slightly more intense $M^{+\bullet}$ than alkanes, but the effect is minor. In addition to the $[C_nH_{2n+1}]^+$ fragment ions seen with alkanes, alkenes show prominent $[C_nH_{2n}]^{+\bullet}$ and $[C_nH_{2n-1}]^+$ species, as evidenced by the spectrum of dec-1-ene in Figure 6.10a, which shows significant ions at m/z 41, 42, 55, 56, 68, 69, etc., as well as the alkane-like series. Although the position of the double bond in a carbon chain may give rise to subtle quantitative differences between the spectra of isomers, it is usually very difficult to assign this from first principles, due to rearrangements during fragmentation.

Simple **cyclic alkenes** give spectra that are qualitatively similar to cyclic alkanes, although the fragment ion series appear $\Delta m/z$ 2 lower, due to the presence of the double bond, and the intensities of the ions also differ from their alkane equivalents. Comparison of the spectrum of cyclohexene (Figure 6.10b) with that of cyclohexane (Figure 6.9b) illustrates these similarities and differences. Both spectra show loss of ethene (to give m/z 54 and 56, respectively) and—perhaps surprisingly—loss of a methyl radical (to give m/z 67 and 69, respectively). This latter observation underlies the importance of hydrogen radical rearrangements in EI fragmentation pathways.

Alkynes, as might be expected by analogy with alkanes and alkenes, can exhibit fragment ion series at $[C_nH_{2n-3}]^+$, although these tend to be in the higher m/z region of the spectrum, with $[C_nH_{2n-1}]^+$ more dominant at lower m/z. The intensity of the molecular ion often varies with the position of the triple bond. Terminal alkynes usually show weak molecular ions, whereas internal alkynes can possess rather abundant molecular ions.

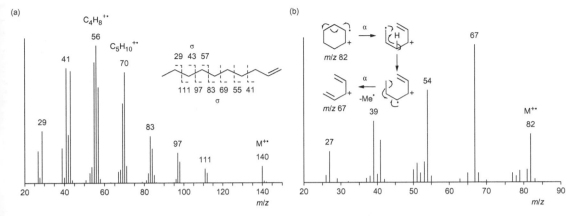

Figure 6.10 EI-mass spectra of (a) dec-1-ene and (b) cyclohexene.

Aromatic hydrocarbons usually show strong molecular ions due to stabilization of the radical cation following removal of a π-electron. The phenyl cation (m/z 77) is often seen, and the ring itself can fragment through loss of ethyne to give m/z 51. If a benzylic group is present, the spectrum is often dominated by the very characteristic tolyl cation at m/z 91 (see Figure 6.11a).

Organic halides exhibit a range of characteristics, depending upon the halogen. Chlorine and bromine display diagnostic isotopic patterns ($^{35}Cl:^{37}Cl = 3.1:1$; $^{79}Br:^{81}Br = 1.03:1$), whereas fluorine and iodine possess essentially single isotopes. Dissociation of chlorides, bromides, and iodides is driven by direct charge-mediated fragmentation to lose the halide radical, whereas fluorides tend to lose HF by initial H-radical migration followed by charge-mediated fragmentation (see Figure 6.11b).

The fragmentation of **alcohols** is largely driven by the oxygen atom, which tends to become the charge site through loss of one of its non-bonding electrons. The molecular ion of aliphatic alcohols is often very weak or absent, and the ion of highest m/z is frequently $[M-H_2O]^{+\bullet}$. α-cleavage around the oxygen-centred radical is usually pronounced, which can reveal the position of the hydroxyl group in a chain (see Figure 6.12a). Phenols generally show much more intense molecular ions, with rather weak $[M-H_2O]^{+\bullet}$ ions.

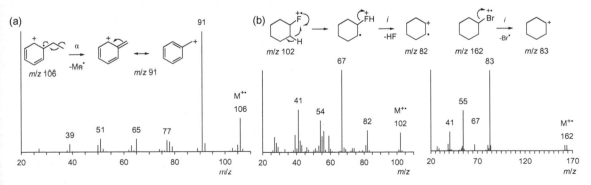

Figure 6.11 EI-mass spectra of (a) ethylbenzene and (b) fluorocyclohexane and bromocyclohexane.

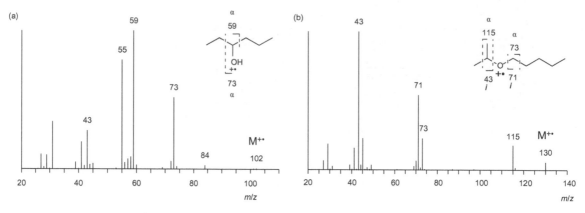

Figure 6.12 EI-mass spectra of (a) hexan-3-ol and (b) isopropyl pentyl ether.

Ethers, despite being isomeric with alcohols, tend to show quite different fragmentation patterns. The molecular ion is generally rather stronger than that for alcohols, with usually no $[M-H_2O]^{+\bullet}$, and whilst α-cleavage is seen, charge-mediated fragmentation is often more significant (see Figure 6.12b).

Thiols and **thioethers** exhibit fragmentation analogous to alcohols and ethers. Even if accurate mass measurement is not available, the presence of sulfur can often be inferred from the ^{34}S isotope signal at m/z M+2 (4.4% of the ^{32}S signal).

The fragmentation of **amines** is dominated by α-cleavage around the nitrogen atom, whether primary, secondary, or tertiary amines.

Aldehydes and **ketones** usually give weak–medium molecular ions (stronger for aromatic compounds). Fragmentation is dominated by α-cleavage around the carbonyl group and, where it can occur, the McLafferty rearrangement. These ions provide characteristic information that can allow complete elucidation of the structure of simple aliphatic and aromatic ketones and aldehydes. Although the EE^+ acylium ions produced by α-cleavage of aliphatic ketones have the same nominal masses as alkyl fragments (m/z 43, 57, 71, 85, etc.), the simple 'clean' pattern they produce is diagnostic (see Figure 6.1). For aldehydes, the loss of H$^\bullet$ (−1) by α-cleavage is distinctive (see Figure 6.13a). Loss of CO by charge-mediated fragmentation often follows initial α-cleavage in both aldehydes and ketones. This interpretation is assisted by the $OE^{+\bullet}$ ions produced by the McLafferty rearrangement. For aldehydes, the McLafferty ion occurs at m/z 44, for 2-ketones at 58, for 3-ketones at 72 (Figure 6.1), etc.

Carboxylic acids, **esters**, and **amides** also exhibit fragmentation dominated by α-cleavage around the carbonyl, and have their own characteristic 'McLafferty ions'. Loss of methoxy and ethoxy radicals from methyl and ethyl esters, for example, are diagnostic, as are the 'McLafferty ions' (where they can form) at m/z 74 and 88, respectively (see Figure 6.13b). Ethyl and longer esters are able to undergo McLafferty rearrangement on this side of the functional group, resulting in the loss of the corresponding alkene. Aliphatic carboxylic acids often give

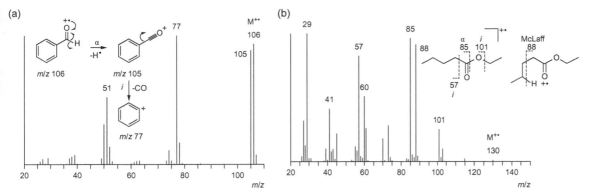

Figure 6.13 EI-mass spectra of (a) benzaldehyde and (b) ethyl pentanoate.

a 'McLafferty ion' at m/z 60, whilst aromatic carboxylic acids show loss of the hydroxyl group as a radical followed by loss of CO. Aliphatic amides can give a 'McLafferty ion' at m/z 59, and their fragmentation patterns are often similar to analogous carboxylic acids and esters.

6.2.6 Mass spectrometry library searching

As mentioned in Chapter 2, the highly reproducible nature of EI mass spectra, acquired under standard conditions, produces a reliable 'fingerprint' for the molecule. A given compound not only exhibits characteristic fragment ions, but does so with consistent relative intensity ratios. This feature makes EI mass spectra ideal for automated searching against libraries of standards. The National Institute of Standards and Technology (NIST) library, for example, contains the spectra of more than 250,000 compounds. Software can allow the user to submit acquired spectra for searching against the NIST or similar libraries, using a matching algorithm. Whilst hits are usually returned with a numerical fitting score between the query and reference spectra, manual comparison is often one of the most reliable methods for confirming a match. Identification is then usually confirmed by sourcing a standard sample of the compound (either commercially or through synthesis) and measuring it on the spectrometer. If the data are generated from GC-MS, then comparison of GC retention times provides additional evidence.

6.3 Interpretation of mass spectra derived from protonation or deprotonation of the sample molecule

Ionization techniques that generate [M+H]$^+$ or [M−H]$^-$, such as ESI, MALDI, or CI, often induce little fragmentation. This property has the clear benefit of revealing the molecule's mass and, if accurate mass measurement is employed, elemental composition. Nevertheless, it is often desirable to induce some fragmentation to provide structural information. As we have

In-source fragmentation. If MS/MS is unavailable it is still possible to induce ion dissociation in API-based instruments by application of an accelerating voltage in the 'transfer region' of the spectrometer between the atmospheric pressure source and the high vacuum section where the analyser is housed. This voltage can be adjusted on most commercial API mass spectrometers, and, if the value is increased sufficiently, collision-induced dissociation usually results. It is often referred to as in-source fragmentation.

seen in the previous chapter, this can be achieved by MS/MS measurement utilizing precursor ion isolation and activation, followed by mass analysis of the product ions.

Next we will address how we can interpret the fragmentation of protonated molecules ([M+H]$^+$) and deprotonated molecules ([M−H]$^-$). We will begin with a brief look at small molecules, before examining larger biomolecules, such as peptides.

6.3.1 Small molecules

As shown in Chapter 2, in positive ion mode, small molecules analysed by ESI, MALDI, or CI usually give rise to protonated molecules ([M+H]$^+$), although it is also common to see cationized species (e.g. in ESI [M+NH$_4$]$^+$, [M+Na]$^+$). The proton is usually found on the most basic site/functional group. If the protonated molecule is activated by CID, the proton may migrate to a less basic position producing a fragile ion which can then fragment by loss of a neutral molecule (see Figure 6.14). Sodiated ions usually exhibit multiple coordination of the metal centre and are generally more stable than a protonated version of the same molecule. Where the sample molecule is already ionic, such as for quaternary ammonium ions ($^+$NR$_4$), then ESI, MALDI, and CI lead to detection of [M]$^+$ without protonation.

Consider the fragmentation of the antidepressant drug paroxetine as an example. The CID MS/MS spectrum of protonated paroxetine (produced by ESI) is shown in Figure 6.15. Fragmentation around the ether functionality dominates. The principal protonation site is the basic piperidine nitrogen, but under conditions of ion activation, proton transfer to the ether oxygen can occur, giving rise to a less stable species that fragments via loss of the neutral alcohol. This gives rise to the base peak at m/z 192 with retention of the charge on the fragment with the highest PA.

Dissociation of deprotonated molecules is usually induced by proton transfer from an acidic residue, by α-elimination (e.g. loss of CO$_2$ from RCO$_2$$^-$), or by hydride transfer (e.g. from the alpha carbon of an alkoxide ion, see Figure 6.16).

The MS/MS fragmentation of deprotonated 2-methoxybenzoic acid is shown in Figure 6.17. It exhibits the loss of both CO$_2$ and formaldehyde by α-elimination.

Protonated and deprotonated molecules [M+H]$^+$ or [M−H]$^-$, as well as cationized (e.g. [M+Na]$^+$) and anionized species (e.g. [M+Cl]$^-$), are even electron ions EE$^+$. When activated by thermal processes such as CID they almost always dissociate by loss of neutral molecules. The absence of an unpaired electron in the ion means that, unlike EI fragmentation, homolytic bond fission is very rare.

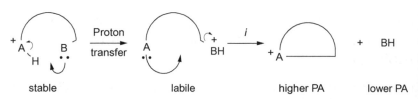

Figure 6.14 Proton migration from a more basic site (A) to a less basic site (B), under ion activating conditions, followed by charge-mediated fragmentation to yield a fragment ion by loss of a neutral molecule BH. The fragment with higher basicity/proton affinity (PA) tends to retain the charge.

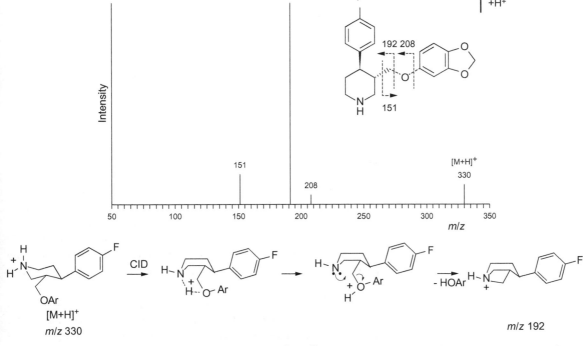

Figure 6.15 MS/MS fragmentation of the protonated molecule ([M+H]$^+$) of paroxetine showing the cleavage sites giving rise to the product ions and a postulated mechanism for the origin of the base peak at *m/z* 192.

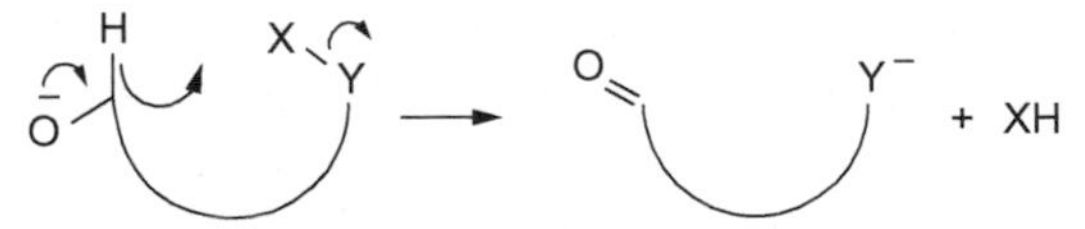

Figure 6.16 Hydride transfer during dissociation of a deprotonated molecule [M−H]$^-$.

m/z 151   *m/z* 107   *m/z* 77

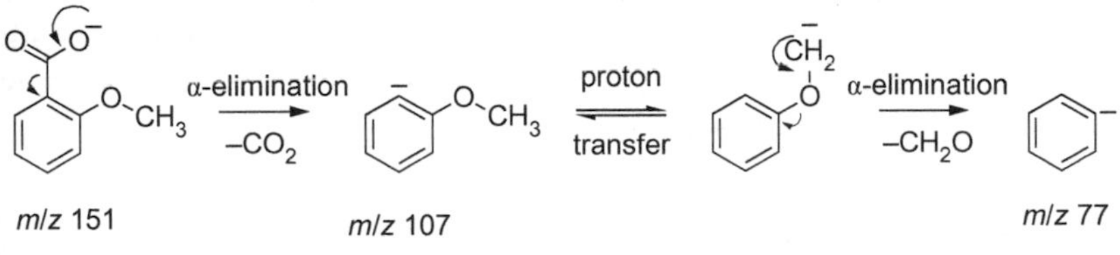

Figure 6.17 MS/MS fragmentation of the deprotonated molecule ([M−H]$^-$) of 2-methoxybenzoic acid showing the sequential α-elimination of CO_2 and formaldehyde.

Although some libraries of MS/MS data from protonated and deprotonated molecules do exist, and can be useful for identification of unknowns, library searching of these spectra is not as reliable as it is for EI-MS. The dependency of MS/MS fragment ion intensities on collision gas and pressure, as well as collision energy, means that ion intensity ratios vary with instrument type and conditions.

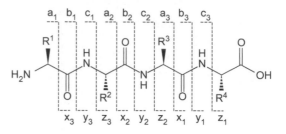

Figure 6.18 The main cleavage sites of peptide MS/MS fragmentation.

6.3.2 Peptides

A major application for the MS/MS fragmentation of protonated molecules (often multiply-protonated molecules produced by ESI) is in the analysis of peptides. Interpretation, or automated database searching of these spectra, allows the peptide's sequence to be determined. We will see the importance of this for protein identification and proteomics in Chapter 8. For the moment we will restrict our discussion to the interpretation of peptide MS/MS spectra.

Figure 6.18 shows the major fragment ion types associated with peptide ion dissociation. Ions originating from the N-terminus of the peptide are denoted by a, b, and c, whilst those from the C-terminus are x, y, and z. A subscript number is applied to the ion to indicate the number of residues from the site of cleavage to the relevant terminus. In CID MS/MS spectra b and y ions dominate, whereas c and z ions are associated with ECD spectra.

The difference in mass between a pair of adjacent ions of similar type (e.g. $m(y_2)-m(y_1)$) will correspond to the condensed mass of the amino acid at that position. As can be seen in Table 6.3, most of the proteinogenic amino acids have a unique mass, and thus their identity can be established from detection of such a pair of ions. This is the principle of peptide sequencing by mass spectrometry using MS/MS analysis. The amino acids lysine and glutamine have the same nominal mass, but can be distinguished by accurate mass measurement on an FT-MS instrument, or even a well-calibrated TOF. Leucine and isoleucine are structural isomers and are therefore isobaric. They cannot be distinguished by simple mass measurement alone.

The term 'proteinogenic amino acid' refers to a member of the subset of naturally occurring amino acids that are incorporated into proteins. There are twenty proteinogenic amino acids in the standard genetic code; see Table 6.3.

The condensed mass of an amino acid refers to its mass−H_2O, as it would appear in a peptide.

6.3.3 CID of peptides

As mentioned previously, the CID MS/MS spectra of peptides are usually dominated by b and y ions. The mechanism of their formation is shown in Figure 6.19. Initially, protonation occurs at the most basic site of the peptide (or sites in the case of multiply charged ions). As we saw for small molecules (Figure 6.14), under conditions of collisional activation the proton(s) may migrate to other sites where they can induce fragmentation. This is the so-called 'mobile proton model' of peptide ion fragmentation. If protonation occurs at one of the peptide bond nitrogen atoms, it can induce fragmentation via the mechanism shown in

Table 6.3 The condensed residue masses of the twenty common proteinogenic amino acids.

Amino acid	3-letter code	1-letter code	Condensed mass
Alanine	Ala	A	71.037
Arginine	Arg	R	156.101
Asparagine	Asn	N	114.043
Aspartic acid	Asp	D	115.027
Cysteine	Cys	C	103.009
Glutamic acid	Glu	E	129.043
Glutamine	Gln	Q	128.059
Glycine	Gly	G	57.022
Histidine	His	H	137.059
Isoleucine	Ile	I	113.084
Leucine	Leu	L	113.084
Lysine	Lys	K	128.095
Methionine	Met	M	131.040
Phenylalanine	Phe	F	147.068
Proline	Pro	P	97.053
Serine	Ser	S	87.032
Threonine	Thr	T	101.048
Tryptophan	Trp	W	186.079
Tyrosine	Tyr	Y	163.063
Valine	Val	V	99.068

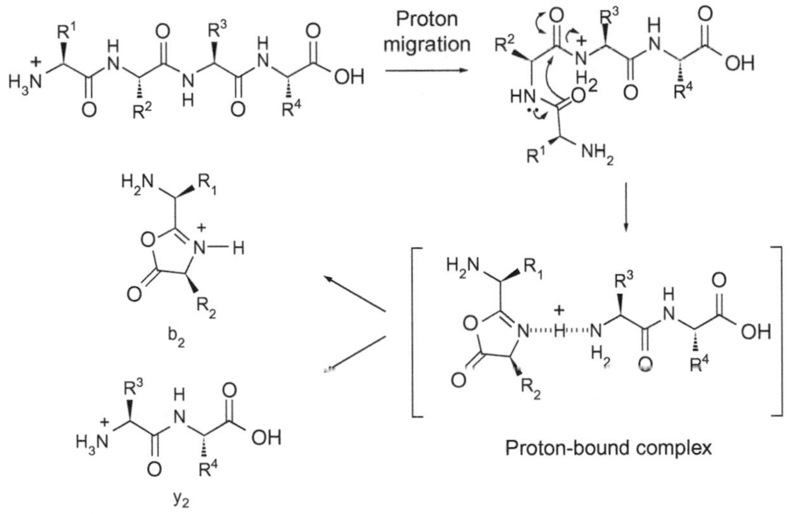

Figure 6.19 The mechanism of CID peptide fragmentation leading to b or y ions. Cleavage of the particular bond shown for the tetra-peptide results in production of the b_2 or y_2 ions.

Figure 6.19. Collapse of the proton-bound complex in the example results in transfer of the proton to either the N- or the C-terminal portion, and leads to generation of b or y ions, respectively. Note that if the precursor ion is multiply charged, then it is possible for both ions to be charged, or for one ion to be multiply charged. There is now significant evidence that the b ions possess an oxazolone ring at their C-terminus, which is indicative of the mechanism presented.

Basic side chains, in particular Arg and Lys and to a lesser extent His, are more likely to carry a charge. During low-energy CID fragmentation of multiply charged peptide ions, regions of the peptide containing these residues are more likely to be seen as ionic fragments, rather than lost as neutrals. If the C-terminal residue is an Arg or Lys, then it is probable that y ions will dominate the MS/MS spectrum. As we will see in Chapter 8, peptides produced by digestion of a protein using the enzyme trypsin always possess a C-terminal Arg or Lys, and so this behaviour is observed.

In addition to the effect of charge location, particular amino acids influence the sites of bond cleavage and hence the distribution of ions detected. Cleavages to the N-terminal side of Pro and to the C-terminal side of Asp, Glu, and His are particularly favoured. The former, proline effect, is due to the gas-phase basicity of the Pro ring and steric effect reducing the likelihood of C-terminal cleavage. Enhanced cleavage at Asp, Glu, and His is due to facile nucleophilic attack of the sidechain functionality at the C-terminal peptide bond.

The low-energy CID MS/MS spectrum of Glu-fibrinopeptide ($[M+2H]^{2+}$) is shown as an example in Figure 6.20. This variant of a peptide involved in

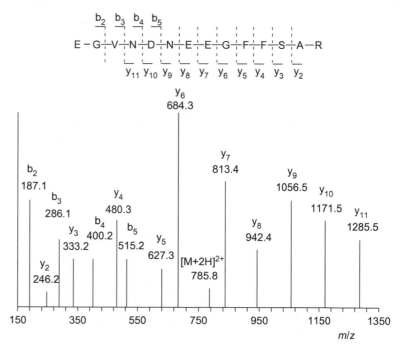

Figure 6.20 The low-energy CID MS/MS spectrum of Glu-fibrinopeptide showing the resulting b and y ions (only the first isotopic peak for each ion is shown for clarity). The insert shows the sequence of the peptide together with the principal fragmentation sites.

blood clotting is commonly used as a standard for optimizing peptide MS/MS conditions, and for calibration. The spectrum is dominated by y ions due to the presence of a strongly basic Arg residue at the C-terminus. The ions y_6, y_7, and y_9 are relatively intense due to preferential cleavage to the C-terminal side of Glu and Asp described previously. The absence of b_1 and y_1 means that incomplete sequence information is provided by the spectrum. This can be overcome, to some extent, by consideration of possible amino acid combinations that correspond to the b_2 ion or the y_2 ion. Notice also that when interpreting spectra manually it is essential to assign b and y ions correctly. This is particularly important in the region of the spectrum where the ion types overlap (i.e. m/z 150–550 in the case of Figure 6.20). Clearly, mass differences between pairs of ions must correspond to known amino acid masses, but sometimes more than one option is possible. In these cases it is helpful to identify a contiguous series of ions that support one of the alternatives. A number of software tools are available to assist *de novo* peptide sequencing.

Where the sequence of a peptide is deduced by interpretation of the spectrum, as opposed to automated database searching, it is referred to as *de novo* sequencing.

6.3.4 **ECD/ETD of peptides**

As mentioned in Chapter 5, electron-capture dissociation (ECD) and electron-transfer dissociation (ETD) are now commonly used for MS/MS of multiply (positively)-charged peptide ions. Unlike CID, fragmentation induced by electron capture or electron transfer tends to produce c and z type ions. A disadvantage of ECD/ETD is that it is relatively inefficient in terms of product ion yield, but it does have the advantages of often giving a more random fragmentation throughout the peptide (increasing sequence information), and, through its non-ergodic nature (see Chapter 5), allowing the mapping of fragile modifications such as phosphorylation (see Chapter 8). The ECD MS/MS spectrum of Glu-fibrinopeptide ($[M+2H]^{2+}$) is shown in Figure 6.21. As with the CID spectrum in Figure 6.20, the presence of a C-terminal Arg residue causes C-terminal ions to dominate, but in an ECD spectrum it is the z rather than y ions that are seen. Another difference between the CID and ECD spectra is the extent of precursor ion fragmentation. Only a small residual signal due to the precursor ion (m/z 785.8) is seen in the CID spectrum, whereas the ECD spectrum exhibits a large unfragmented precursor ion. This is typical of the ECD process, which is relatively inefficient in its yield of fragment ions. The two spectra illustrate the complementarity of CID and ECD; if both techniques are employed full sequence coverage of Glu-fibrinopeptide can be obtained.

6.3.5 **Carbohydrates**

Interpretation of the fragmentation patterns of carbohydrates (usually generated by low-energy CID) can provide important structural insights. Identifying glycan structure from MS/MS spectra is, however, rather more challenging than peptide sequencing. There are several reasons for this, including the isobaric nature of

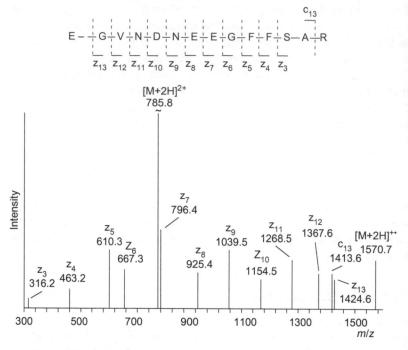

Figure 6.21 The ECD MS/MS spectrum of Glu-fibrinopeptide, showing the resulting c and z ions (only the first isotopic peak for each ion is presented for clarity). The insert shows the sequence of the peptide together with the principal fragmentation sites.

many hexoses (e.g. glucose, galactose, and mannose), the possibility of different linkage patterns across the carbohydrate rings, and the existence of branching in the glycan chain. We will consider here only the very basic aspects to illustrate the principles of oligosaccharide analysis by MS/MS.

As with peptide fragmentation, there is a standard nomenclature for carbohydrate fragment ions. The letters A, B, and C are used to denote ions originating from the end of the glycan chain that does not contain the terminal anomeric carbon, whilst X, Y, and Z represent ions that do include the terminal anomeric carbon, with, in both cases, subscript numbers employed to indicate the number of residues from the end of the chain. B, C, Y, and Z ions all arise from cleavage of glycosidic bonds (see Figure 6.22a), whilst A and X ions are caused by fragmentation through the carbohydrate rings (see Figure 6.22b). Because the rings can cleave by dissociation of different pairs of bonds, additional numbering is used to show which bonds break. This is given as a prefix before either A or X, according to the bond numbering shown in Figure 6.22c.

Carbohydrates are commonly analysed in positive ion mode as protonated or cationated ([M+Na]$^+$ etc.) ions, and in negative ion mode as deprotonated or anionated ions ([M+NO$_3$]$^-$ etc.). Whilst the positive ions usually dissociate to produce B- and Y-type ions, negative ions dissociate to give through-ring cleavage

---

The terminal anomeric carbon of a carbohydrate is the carbon atom within the final hemiacetal functional group. It is usually the point of attachment to a protein through an N- or O-link. In free carbohydrates this is also referred to as the reducing end.

Capital letters are used for carbohydrate fragment ions to differentiate them from peptide fragment ions.

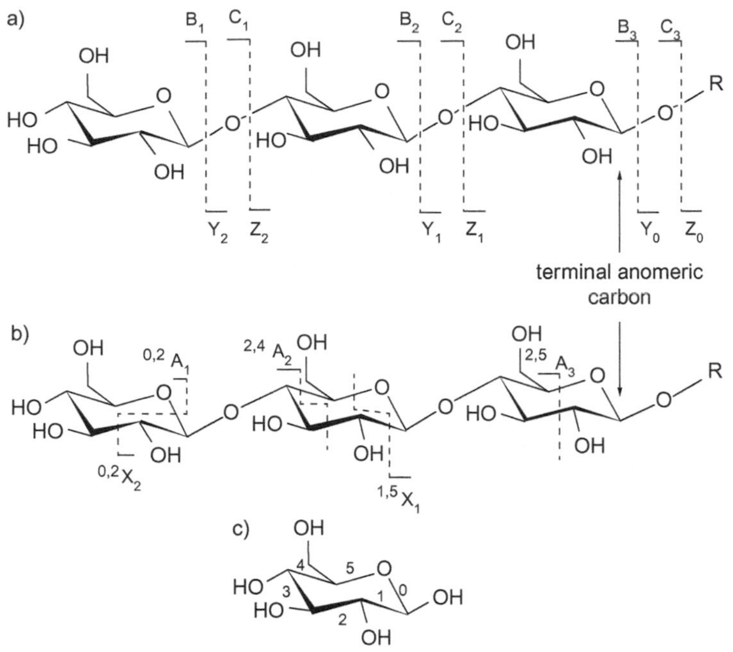

Figure 6.22 Nomenclature for the MS/MS fragmentation of carbohydrates. For simplicity, a 1,4-β-linkage pattern of glucose is shown, but many other linkages and combinations of monosaccharides are possible, which often makes the interpretation of glycan spectra difficult. a) glycosidic bond cleavages, b) through-ring cleavages, and c) numbering of bonds in the carbohydrate ring to specify through-ring cleavage patterns.

products. These can be particularly useful in identifying linkage types, as the mass of the fragment ion changes with the branching pattern.

Much of the interest in complex carbohydrate analysis by MS is in the context of glycoproteins. Glycans are commonly composed of a mixture of 'simple' monosaccharide residues (e.g. galactose (Gal), mannose (Man), fucose (Fuc), and glucose (Glc)), as well as acetylamino-monosaccharides such as N-acetyl-glucosamine (GlcNAc). A very common motif is the 'core' pentasaccharide, which has the sequence $Man_3GlcNAc_2$-Asn (i.e. attached to the Asn residue of a protein through an N-link). This can provide a starting point for interpretation of glycan structure, but the possible additions to this core are numerous and complex.

6.3.6 Oligonucleotides

Whilst mass spectrometry is commonly used in the characterization of synthetic oligonucleotides, its use in routine sequencing of DNA and RNA of biological origin is currently much less widespread. This is principally because of the success of Sanger-type sequencing methods, which are capable of sequencing long strands of oligonucleotides with high sensitivity. Mass spectrometry (especially

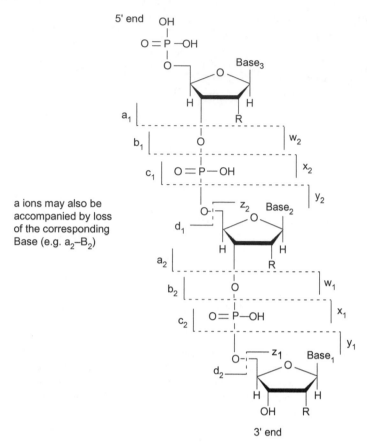

a ions may also be
accompanied by loss
of the corresponding
Base (e.g. $a_2 - B_2$)

Figure 6.23 Nomenclature for the MS/MS fragmentation of oligonucleotides.

MALDI-MS) can be used to detect the products of these types of sequencing as an alternative to the more common electrophoresis-based methods. CID MS/MS of oligonucleotides does yield a series of fragment ions from which sequence information can be extracted. In practice, however, MS/MS sequencing is generally limited to short oligonucleotides (fewer than twenty bases), and, due to the presence of strongly acidic phosphate residues, is often performed on negative ions ($[M-nH]^{n-}$). The typical fragmentation modes of oligonucleotides are shown in Figure 6.23.

An area where MS is showing significant promise is in the detection of single nucleotide polymorphisms (SNPs). These are single base pair mutations at a specific position in the chromosome. Molecular biology techniques are employed to extend a short oligonucleotide primer complementary to a portion of the gene of interest, and MALDI-MS is used to measure the mass of the resulting extended primer. Genetic variation is apparent by the change in mass detected by the presence of a different base.

6.4 Summary

From the material presented in this chapter you should be familiar with the following:

- The principal mechanisms of dissociation of molecular ions in EI mass spectra.
- The fragmentation of common functional groups/types of organic molecules in EI mass spectra.
- The basics of CID fragmentation of protonated and deprotonated small molecules.
- The fragmentation of peptides by CID and ECD.
- The basics of fragmentation of carbohydrates and oligonucleotides by CID.

6.5 Exercises

6.1 Identify the most likely charge site(s) on the following molecules under EI conditions.

6.2 Identify which of the following ionic masses (and formulae) are even electron ions and which are odd electron ions.

m/z 86 ($C_5H_{10}O$) m/z 93 (C_6H_7N) m/z 61 (C_3H_6F) m/z 158 ($C_{10}H_{10}N_2$)

6.3 Predict the fragment ions and neutrals arising from charge mediated fragmentation of the following ions. Calculate the expected m/z of each ionic product.

6.4 Predict the fragment ions and neutrals arising from α-cleavage of the following molecular ions. Calculate the expected m/z of each ionic product (more than one product ion may be possible in each case).

6.5 Predict the fragment ions and neutrals arising from McLafferty rearrangement of the following molecular ions. Calculate the expected m/z of each ionic product (more than one product ion may be possible in each case).

6.6 Interpret the EI spectrum of heptan-4-ol shown in the illustration by assigning structures to the fragment ions observed.

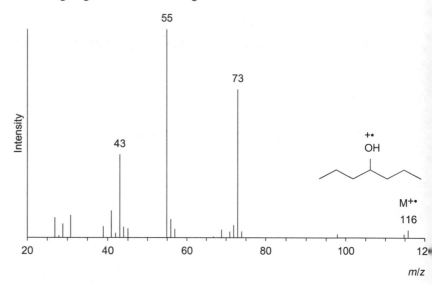

6.7 Interpret the EI spectrum shown in the illustration and suggest a structure for the unknown compound. Use the data in Tables 6.1 and 6.2 to help you.

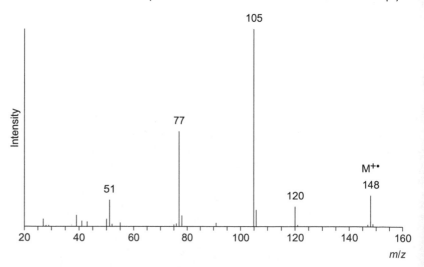

6.8 The protonated molecule ($[M+H]^+ = m/z$ 455.2) of methotrexate fragments under CID MS/MS conditions to give the major product ions m/z 308.1, 175.1, and 134.1. Suggest structures for these ions and propose mechanisms to explain their formation.

Methotrexate

6.9 CID MS/MS fragmentation of a doubly-charged peptide precursor ion at m/z 600.25 gave rise to the following singly-charged b and y ions (note that a full set of ions was not detected). Using the condensed masses of the amino acids provided in Table 6.3, determine the sequence of the peptide (from N to C termini). Hint: to check your answer, compare the sum of the condensed masses of the amino acids in your sequence (plus the mass of one water molecule) with measured mass provided.

b1	132.048	y4	510.216
b2	203.085	y5	639.259
b3	290.117	y6	736.312
b4	377.149	y7	823.344
b5	464.181	y8	910.376
b6	561.234	y9	997.408
b7	690.276	y10	1068.445
b8	793.286		
b9	894.333		
b10	1050.434		

6.10 ECD MS/MS fragmentation of a doubly-charged peptide precursor ion at m/z 810.32 gave rise to the following singly-charged c and z ions (note that a full set of ions was not detected). Using the condensed masses of the amino acids provided in Table 6.3, determine the sequence of the peptide (from N to C termini). Hint: to check your answer, compare the sum of the condensed masses of the amino acids in your sequence (plus the mass of one water molecule) with measured mass provided.

c1	115.087	z1	131.035
c2	244.129	z2	234.044
c3	341.182	z3	348.087
c4	442.230	z4	477.130
c5	555.314	z5	605.188
c6	670.341	z6	734.231
c7	799.383	z7	821.263
c8	886.415		
c9	1015.458		
c10	1143.517		

6.6 Further reading

Cole, R. B. (ed.) (2010). *Electrospray and MALDI Mass Spectrometry*. New York: Wiley.

Demarque, D. P. et al. (2016). 'Fragmentation reactions using electrospray ionization mass spectrometry: an important tool for the structural elucidation and characterization of synthetic and natural products', *Nat. Prod. Rep. 3*, 432–55.

McLafferty, F. W. and Tureček, F. (1993). *Interpretation of Mass Spectra*, 4th edn. California: University Science Books.

Watson, J. T. and Sparkman, O. D. (2007). *Introduction to Mass Spectrometry*, 4th edn. Chichester: Wiley.

7 Separation techniques and quantification

7.1 Introduction

The coupling of separation techniques with mass spectrometers has become a ubiquitous part of modern mass spectrometry. Gas chromatography, liquid chromatography, electrophoresis, and ion mobility devices can each be combined with a mass spectrometer to form an integrated system. These so-called 'hyphenated techniques' can enable more selective and comprehensive analysis of complex mixtures, as well as provide a platform for quantitative analysis of individual analytes.

In this chapter the basic principles of analyte separation will be addressed in the context of mass spectrometry. This will include separations that occur inside the mass spectrometer before the mass analyser, as well as those that take place prior to ionization using interfaced techniques. We will discuss how combined separation and detection systems are connected, and how the analysis is combined to provide additional chemical and quantitative information. We will pay particular attention to chromatographic techniques, but detailed chromatographic theory is beyond the scope of this book and will only be introduced at a very basic level, or when it has direct relevance to mass spectrometry performance. The principles of quantitative analysis using chromatography coupled to mass spectrometry will be outlined and the advantages and disadvantages of different types of chromatographic calibration discussed.

'Calibration', in the context of quantification by mass spectrometry, refers to calibration of peak areas with respect to concentration, and not the accuracy of m/z measurements. It should therefore not be confused with mass analyser calibrations used to improve mass accuracy (discussed in Chapter 4).

7.2 Why couple separation techniques to mass spectrometry?

The signal response from a mass spectrometer is the ion abundance associated with specific m/z values. This ion abundance is related to the concentration of sample molecules or atoms being analysed, but is also affected by a number of additional variables. For example, the type of ionization source and ionization mechanisms being used (Chapter 2), the molecular structure of the compound

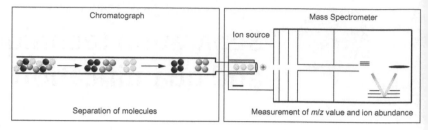

Figure 7.1 A schematic showing the generic coupling of a chromatograph with a mass spectrometer. Compounds in a sample are separated chromatographically based on their individual interactions with the mobile and stationary phases, and then ionized and detected sequentially by the mass spectrometer in a continuous process.

being ionized, the mass and charge state, and whether other compounds are present in the sample. The difficulty in controlling these variables makes it difficult to use mass spectrometry alone as a quantitative tool. However, coupling separation techniques such as chromatography with mass spectrometry can be a very effective strategy for producing highly sensitive and quantitative analyses. Figure 7.1 shows a schematic of a generic chromatography system coupled with a mass spectrometer.

7.2.1 Matrix effects and ion suppression

Matrix effects are interference in the ion signal of a particular analyte caused by the presence of other compounds. They are often associated with the analysis of biological or environmental samples, although exogenous contaminants (such as plasticizers from samples coming in contact with plastic-ware) can lead to matrix effects in the analysis of otherwise 'purified' compounds. Matrix effects often reduce (but can occasionally enhance) ionization efficiency. When the analyte signal is attenuated this is often referred to as 'ion suppression'—a process which can have a significant negative affect on sensitivity and confound the ability to quantify analyte concentrations.

The mechanism of ion suppression is not well understood, but is associated with samples of increased chemical complexity, basicity or acidity, the concentration of the analyte itself, and the presence of non-volatile compounds—in particular, salts. It is important to note that ion suppression is a property of the ionization process and is not reduced by post-source strategies designed to increase selectivity and sensitivity such as SIM, SRM, and MRM (Chapter 5). Approaches which help to remove charge competition and reduce signal loss during ionization are of particular importance in mitigating against the unwanted effects of ion suppression. Chromatographic separation prior to mass spectrometric analysis, is one of the most effective ways to remove ion suppression, although it does not guarantee elimination of its effects in all cases.

7.3 Chromatography coupled to mass spectrometry

When chromatography and mass spectrometry are combined they can provide a number of advantages that go beyond the capabilities of either technique used separately. These include:

1) Measurement of time-resolved mass spectra for multiple compounds in a sample.

2) A straightforward/automatable approach to introduce samples into a mass spectrometer.

3) A way to analyse complex mixtures by reducing matrix effects and ion suppression.

4) A technique to distinguish structural and stereo isomers.

5) A method to combine relative and absolute quantitation with analytical mass spectrometry.

There is a lack of consensus in the literature on the appropriate way to indicate coupling of chromatography and mass spectrometry. Should it be LC/MS, LC-MS, LCMS, or LC MS? Exactly the same question can be asked of gas chromatography coupled to mass spectrometry. All these acronyms can be found in contemporary scientific literature, and all refer to the same thing: the coupling of mass spectrometry and the particular form of chromatography indicated. Here we use the hyphenated form (LC-MS and GC-MS) in accordance with 2013 IUPAC recommendations and commensurate with the often used term 'hyphenated technique'.

7.3.1 Time-resolved mass spectra

Irrespective of the type of ionization source or analyser used, when chromatography is coupled to mass spectrometry the data collected have an additional time-dimension (Figure 7.2) as the chromatographic process separates compounds in time. How long it takes to do this can vary from less than a minute to more than an hour, depending on the chromatography method used and the analytes of interest. However, chromatographic separation of analytes is a much slower process than subsequent analysis by mass spectrometry.

Instead of a static mass spectrum being produced across the data acquisition period (as is the case for analysis of a sample directly infused into an ion source), chromatographic coupling leads to a dynamic series of spectra where the presence of ions and their abundance changes with time. For every 'scan' or 'packet' of ions processed by the mass analyser, a spectrum is recorded, and there can be many scans, across a designated mass range, per second (see Chapter 3). Mass spectra therefore record the changes in the abundance of ions over time as they elute from the chromatograph. A total ion chromatogram (or total ion current chromatogram) (TIC) represents the total ion current detected (across the entire mass range measured) as a function of chromatographic retention time. Figure 7.2 provides an example of a total ion chromatogram from the analysis of a mixture of 170 compounds analysed by ion-exchange chromatography coupled directly to a mass spectrometer. Selected single spectra at four different time points show the changes in the mass spectrum over time.

A TIC is analogous to a UV chromatogram where UV absorbance of eluting compounds is plotted over time in some types of chromatography system. The difference for mass spectrometry detection is that the TIC presents changes in total ion abundance rather than UV absorption.

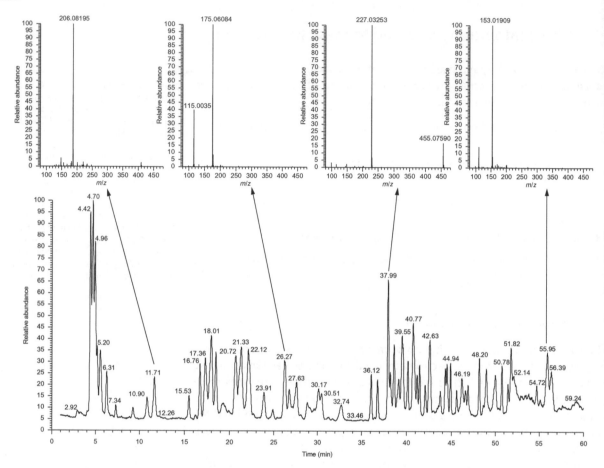

Figure 7.2 A total ion chromatogram showing the change in total ion abundance (vertical axis) over the duration of the 60-minute (horizontal axis) LC-MS run.

Mass spectra and total ion chromatograms are simply two ways of displaying the ion abundance data collected by the mass spectrometer over time. It is possible to interconvert between a mass spectrum and a mass chromatogram display. For example, a mass spectrum can be generated at any point in time from the *total ion chromatogram*, and conversely an *extracted ion chromatogram* (EIC) can be generated for a specific *m/z* value or wider *m/z* range. The EIC for a particular *m/z* value is created by plotting its ion abundance as a function of retention time. This is a powerful way of displaying the chromatographic peaks for selected analytes in a mixture. These fundamental capabilities of displaying ions, and their changes in abundance over time, are now built into all data analysis software associated with modern mass spectrometers.

Figure 7.3 shows how the mass spectrum and extracted ion chromatogram are related. Figure 7.3a shows the total ion chromatogram, and Figure 7.3b shows the mass spectrum generated at a specific retention time (Rt) of 11.71 min.

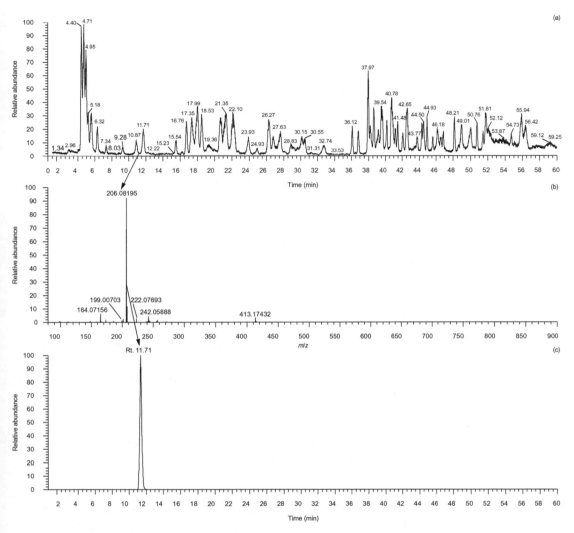

Figure 7.3 (a) shows a total ion chromatogram for the LC-MS analysis of a complex mixture of compounds, (b) shows the mass spectrum at retention time (Rt) 11.71 min where a compound elutes represented by ions at *m/z* value of 206.08195, and (c) shows the extracted ion chromatogram (EIC) associated with the *m/z* value of 206.08195 (5 ppm mass accuracy).

The total ion chromatogram is usually interactive in mass spectrometry software and the TIC can be used to generate a mass spectrum at a particular elution time. In the example in Figure 7.3, a mass spectrum at retention time 11.71 min shows a major mass spectral peak with an *m/z* value of 206.08195. This is the *m/z* value corresponding to the negative ion electrospray analysis of N-acetyl-L-phenyl-alanine ($C_{11}H_{13}NO_3$).

Figure 7.3c shows the EIC generated for *m/z* 206.08195 within a 5 ppm range. The single chromatographic peak from the EIC represents all the deprotonated

In a mass spectrum the peaks are referred to as *mass spectral peaks*, and in a chromatogram the peaks are referred to as *chromatographic peaks*. Both types of peak are separate entities that represent quite different information. When describing either, care should be taken to ensure that they are distinguished explicitly.

N-acetyl-L-phenylalanine ions formed from the N-acetyl-L-phenylalanine molecules in the sample injected, and hence both the peak area and height are proportional to the concentration of this analyte in the sample.

7.3.2 Chromatographic peak area and concentration

The dynamic relationship between the chromatogram and the mass spectrum is one of the reasons why chromatography and mass spectrometry can combine to provide a tool for quantitative analysis. The chromatographic peak area and height associated with a particular analyte, is representative of its concentration. This is because analyte molecules are initially retained by the stationary phase and then 'eluted' at a specific point in time. All the molecules of a particular analyte that interact with the chromatographic column elute within a narrow window of time, subsequently represented by a single peak in the chromatogram with a maximum at the compound's 'retention time'. Ions are formed from these eluting molecules in the mass spectrometer's ion source, and their abundance is proportional to the total amount of analyte present in the injected sample and therefore represents its concentration directly.

The specific retention time of an analyte represents how long it takes to travel from initial injection into the chromatographic system to its elution and arrival at the detector. This retention time is both a function of chemical structure and chromatographic conditions. Many compounds (including a range of structural isomers) will have a unique retention that is reproducible under the same analytical conditions. Thus, the abundance of the same analyte in different samples can be compared directly. The retention time provides selectivity that is orthogonal, and additional to an ion's *m/z* value. Because chemical species are separated chromatographically the effects of ion suppression are minimized, often enhancing sensitivity and reproducibility, even for samples with high analyte complexity.

It is important to note that chromatographic peak area alone does not tell us the absolute amount of an analyte present; the peak area is not in concentration units. However, comparing peak areas for the same analyte does indicate relative similarities or differences in concentration. It is also important to note that this will only be valid within a certain concentration range. Beyond this range (above and below) a mass spectrometer's detector response may not be linear with concentration. For example, if too many ions arrive at the detector at once not all will be detected and the concentration will be underestimated. Below the limit of quantitation, so few ions arrive at the detector that the ion signal is insufficient for accurate integration. Between these extremes a linear relationship exists between detector response and concentration for the majority of ion sources; chromatography systems coupled to mass spectrometers can therefore be used for quantitative analysis. The concentration range across which the response is linear is analyte and sample specific for a particular analytical platform and must be determined experimentally. To ensure reliable quantitation, it is therefore important to calibrate the spectrometer's response across the concentration range of interest using appropriate standards (see Section 7.7).

7.3.3 Scan speed, mass range, and sensitivity

As a result of coupling chromatography with a mass spectrometer the 'scan time' or 'analysis speed' of the mass analyser becomes an important factor (see Chapter 3). The fact that a mass spectrometer has a finite scan time means that there are only a finite number of data points per chromatographic peak (Figure 7.4), and usually a minimum of ten data points are required to ensure that the chromatographic peak can be produced accurately. Note that for a fixed scan time the chromatographic peak width will have a direct impact on the number of data points collected.

In Figure 7.4 the chromatographic peak is relatively wide. Note that each scan provides a mass spectrum, with the total abundance of the analyte ions changing as the peak elutes, and the maximum occurring at the centre of the peak. The sum of these ion abundances across the peak provides the total number of ions representative of the analyte.

Table 7.1 shows a comparison of the typical scan speeds for different types of mass analyser. These can have a direct bearing on analytical performance when a chromatography system is coupled to a mass spectrometer.

Triple quadrupole mass analysers have traditionally been favoured for quantitative analysis. Given their low resolution (see Chapter 3), relatively poor mass accuracy, and average scan speeds (Table 7.1), it is not entirely obvious why this should be. However, two of their advantages are, a wide linear response range and the ability to perform single reaction monitoring (SRM) and multiple reaction monitoring (MRM) experiments (discussed in Chapter 5). SRM avoids the problems associated with scan time by continuously monitoring a selected m/z value that represents the analyte of interest. Because there is no time spent scanning other m/z values, the number of data points across a chromatographic peak is maximized, considerably enhancing sensitivity. The scan time for TOF mass analysers is very fast and these are also commonly coupled

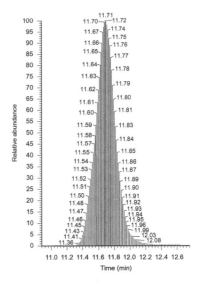

Figure 7.4 An Ilustration of how a number of measurements are made across an eluting peak at a particular m/z value. In this example approximately fifty mass spectral scans are acquired across the single chromatographic peak.

Table 7.1 Typical scan speeds associated with different types of mass analyser.

Analyser type	Typical scan per second
Magnetic sector	0.5–2
Quadrupole	0.1–1
Quadrupole ion trap	0.01–1
Linear ion trap	0.01–1
Time-of-flight	0.1
Orbitrap FTMS	0.1–1
FT-ICR-MS	0.5–10

with chromatography systems, although in general they do have a narrower linear response range than triple quadrupole mass analysers. In conclusion, the importance of scan time clearly depends on experimental requirements, such as how many compounds at different *m/z* values are being analysed in a sample, how wide the chromatographic peaks are, and whether the instrument can perform SIM, SRM, or MRM experiments.

7.4 Liquid chromatography mass spectrometry (LC-MS)

Liquid chromatography is based on differential partitioning of analytes between a mobile phase and a solid stationary phase (housed in a chromatographic column) where the mobile phase is a liquid solvent, or mixture of liquid solvents, with differing polarities. A liquid sample is injected into the mobile phase, which flows across the stationary phase surface. Interactions between analyte molecules and the stationary phase slow passage of the analyte through the chromatographic column.

Liquid chromatography coupled to mass spectrometry (LC-MS) is now probably the most common hyphenated technique in mass spectrometry. However, historically it was difficult to interface efficiently. One of the major hurdles was to couple the LC system directly with the ionization source of the mass spectrometer. This was problematic for high vacuum sources because it required continuous removal of large volumes of eluting solvent. Despite a number of innovative solutions it was not until the advent of atmospheric pressure ionization (API) sources in the 1980s and 1990s that LC-MS coupling was achieved with high efficiency. In particular there are two aspects of electrospray ionization (ESI) that made it an ideal ionization source for LC-MS. First, rapid evaporation of the solvent is an integral part of the ionization process, and second, gas-phase charged molecules are formed from non-volatile compounds at atmospheric pressure. Ultra-high-performance chromatography (UHPLC), capillary, and nano-liquid chromatography use lower flow rates than conventional HPLC and are more efficiently coupled with ESI, although modern ESI sources are also capable of interfacing with flow rates above 1 mL per minute. Table 7.2 provides approximate flow rates for different types of liquid chromatography system that are coupled directly with mass spectrometers.

7.4.1 Types of LC-MS

A single liquid chromatography technique does not provide a universal analysis platform. There are different LC approaches and different stationary phases available that are selective for specific types of chemical interaction. This inevitably leads to better performance for certain types of analyte with particular polarities or functional groups present. For example, in *reversed-phase-liquid chromatography* (RP-LC) the mechanism of analyte retention is based on partitioning of analyte

The void volume is the volume of the mobile phase in a chromatographic column and all the tubing to and from the column to the mass spectrometer (V_m or V_0). This is determined by the column inner dimensions and the proportion of space the stationary phase occupies. The period of time it takes to pump the void volume through the column to the detector is the time it takes for compounds that are not retained by the chromatographic system to arrive at the detector from the point of injection. This is usually indicated by the first peak at the start of a chromatogram.

Table 7.2 The flow rates of the most common types of liquid chromatography systems coupled with mass spectrometers.

Liquid chromatography system	Abbrev.	Typical column ID (mm)	Typical flow rates (μL/min)
High-performance liquid chromatography	HPLC	2.1–4.6	100–1000
Ultra-high-performance liquid chromatography	UHPLC	1–2.1	100–400
Capillary liquid chromatography	Cap-LC	0.3–0.5	5–50
Nano-liquid chromatography	Nano-LC	0.1	<1

molecules between a polar eluent and non-polar alkyl chains bonded directly to the surface of silica particles in the column. The degree of retention therefore depends on the relative hydrophobicity of the analyte. Compounds are retained under aqueous conditions and then elute at higher concentrations of organic solvent (most commonly methanol or acetonitrile). Highly polar and ionic compounds will not be retained using reversed-phase chromatography because they are not hydrophobic enough to interact significantly with the stationary phase under aqueous conditions. Such compounds elute simultaneously in the 'void volume' without being retained at all.

For retention of analytes that are already ions in solution (anions, cations, zwitterions), ion-exchange chromatography (IEC) can be a more suitable chromatographic technique. Here a resin-based stationary phase is modified with permanently ionized functional groups, which provide a surface for electrostatic interactions with analyte ions and highly polar molecules. A higher concentration of ions of the same polarity as the analyte in the mobile phase is introduced, usually by way of a concentration gradient with, for example, OH^- or H^+ using anion and cation exchange chromatography respectively. The mobile phase ions displace retained analyte ions at different times according their strength of binding to the stationary phase. The mixture of aqueous and/or organic solvents used in RP-LC are well matched to the requirements of ESI, but the high ion content associated with OH^- or H^+ and their counter ions, which are commonly used in ion-exchange chromatography, are generally incompatible with ESI. Subsequently IEC has not been directly coupled with MS until relatively recently. For efficient coupling, a mechanism of removing high concentrations of mobile phase ions is required. Electrochemical suppression systems that remove high concentrations of OH^- or H^+ (for anion and cation exchange chromatography respectively), converting these ions to H_2O in the eluting mobile phase, have been developed and have enabled the use of online IEC-MS techniques (see Figure 7.5).

Table 7.3 lists a range of chromatographic techniques and their suitability for coupling to mass spectrometry. It should be noted that as well as matching the type of chromatography to the analyte for effective separation, the mobile phase also needs to be compatible with the mass spectrometer's ionization method to ensure functionality and adequate sensitivity.

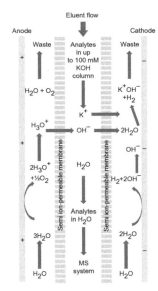

Figure 7.5 Illustration of an electrochemical suppression cell used in anion-exchange chromatography to enable coupling with electrospray ionization mass spectrometry. This approach uses a strong hydroxide ion gradient to elute ions from the column. The electrochemical cell removes cations and hydroxide ions from the mobile phase after chromatographic separation but before the eluent enters the mass spectrometer.

Table 7.3 Mechanisms of retention used in liquid chromatography, the types of compounds for which they are suitable, and notes on their compatibility with mass spectrometry.

Liquid chromatography separation mechanism	Acronym	Retention mechanism	Types of analyte	Suitability for MS detection
Reversed phase	RP-LC	Hydrophobic interactions	Medium to high polarity compounds	Very compatible
Ion exchange	IEC	Ionic interactions	Ionic and highly polar compounds	Not generally compatible unless ion suppression used
Ion pairing	IP-LC	Ion-pair formation and hydrophobic interactions	Medium to highly polar compounds	Use with caution IP reagents can lead to ion suppression and contamination
Hydrophilic interaction	HILIC	Hydrophilic and polar interactions	Medium to highly polar compounds	Very compatible
Mixed-mode	MM-LC	Hydrophobic, polar and ionic interactions	Low to highly polar compounds	Can be suitable depending on the pK_a of the embedded groups
Normal phase	NP-LC	Hydrophilic interactions	Low to medium polarity compounds	Not generally suitable

7.4.2 UHPLC-MS and nano-LC-MS

Ultra-high-performance liquid chromatography (UHPLC) was developed in 2004 by Waters Corporation. It differs from HPLC by utilizing sub-2 μm silica particles in columns based on reversed-phase stationary phases leading to higher system backpressures than for conventional HPLC. UHPLC flow paths and pumping systems are generally designed for working pressures up to 1000 bar (although systems have recently been developed with operating pressures up to 1400 bar and 1.6 μm stationary phase silica particles). These compare with a traditional HPLC operating pressure of around 400 bar and traditionally 5 μm silica particles. The benefits of smaller particles are an increase in efficiency (narrower, higher-resolution chromatographic peaks), and slower optimal flow rates. Both tend to be beneficial for coupling with mass spectrometry.

Nano-electrospray (nano-ESI) is one of the most efficient and sensitive ionization methods available for coupling with mass spectrometry (Chapter 3). Its compatibility with nano-liquid chromatography (nano-LC) means the latter requires very little sample volume and is extremely efficient in its use of mobile phases. However, due to the low flow rates, narrow flow path, and micro-unions it is much slower to perform an equivalent chromatographic separation compared with HPLC or UHPLC, and is generally less robust for routine use. It is more challenging to keep retention times consistent for the same analyte, for example. It is used commonly in proteomics (see Section 8.4) but less often for the quantitative analysis of small molecules.

7.4.3 LC-MS challenges

Although there are numerous benefits in coupling LC and MS systems, a number of inherent compromises are required that can influence the performance of both systems. For example, mobile phase additives can have significant enhancing effect on the chromatographic retention time and peak shape for certain analytes. Trifluoroacetic acid (TFA) is a common additive used in chromatography that can provide chromatographic enhancements when added at a few percent by volume or less to the mobile phase. However, its presence leads to large, non-analyte peaks in the MS spectrum. TFA can also be difficult to remove from surfaces, leading to contamination of the LC system and mass spectrometer ion source. TFA is also well known for promoting subsequent signal suppression, particularly in positive ion mode ESI, by formation of gas-phase ion-pairs with analyte ions.

Inorganic buffers such as potassium phosphate have traditionally been used in LC to control solvent pH, and thus ionization state of analytes, but with ESI especially, these produce large background signals, lead to ion suppression, and can coat the inside of the ion source with residue. To avoid these effects, in general only volatile buffers are recommended for LC-MS, but these are not always as effective in controlling chromatographic conditions (see Table 7.4).

Table 7.4 Common buffers and additives used in HPLC, their pH range, and their suitability for LC-MS.

Volatile buffer/additive	pH range	Suitability of LC-MS
Ammonium formate	2.7–4.7	Yes
Ammonium acetate	3.6–5.6	Yes
Acetic acid	4.3–5.3	Yes
Formic acid	2.7–4.7	Yes
Ammonium bicarbonate	6.5–8.5	Yes
Trifluoroacetic acid	2	Not recommended
Boric acid	8–10	Yes
Ammonia	11.6	Yes
Ammonium carbonate	6–7	Yes
Pyrrolidine	10.3–12.3	Yes
Ammonium hydroxide	8.3–10.3	Yes
Ammonium bicarbonate	6.7–11.3	Yes
Triethylamine	9.7–11.7	Yes
Tributylamine	N/A	Yes
TRIS	7–9	No
Phosphoric acid	<1	No
Phosphate (pK1)	1–3	No
Phosphate (pK2)	6–8	No
Phosphate (pK3)	11–13	No

The addition of small amounts of acetic or formic acid in the mobile phase (in the region of 0.1%) can help promote the formation of positive ions during the electrospray process. However, a slightly acidic mobile phase may also affect the ionization state of certain analytes, making them more difficult to retain chromatographically. Clearly there are numerous compromises, and one of the skills of the LC-MS analyst is the ability to take these into account when developing optimal methods.

7.4.4 Synthetic chemistry applications for LC-MS

LC-MS-based reaction monitoring in synthetic chemistry can provide a more sensitive and selective alternative to traditional TLC. However, hybrid techniques have also become popular with, for example, direct analysis of TLC plates by DESI, DART, and MALDI mass spectrometry (Chapter 2). These methods tend to have a speed advantage over LC-MS, and retain specificity and selectivity whilst interfacing with the well-established TLC workflow.

LC-MS provides a fast and efficient approach to synthetic reaction monitoring and has a number of advantages over methods such as thin-layer chromatography (TLC). LC-MS lends itself well to confirming the identity of synthesized compounds from accurate mass and retention time data, as well as identifying and characterizing side reactions. A quantitative approach may also be used to determine reaction rate and product yield (see Section 7.7). A rapid reversed-phase LC-MS approach is typically used, with sample injection volumes usually between 2µL and 10µL, and chromatographic run times of ten minutes or less. Many targeted methods suitable for analysis of specific analytes are published in mass spectrometry and subject-specific journals.

7.4.5 Intact protein analysis by LC-MS

The characterization of denatured, intact proteins is particularly amenable to reversed-phase chromatography. The option of C18, C8, and C4 alkyl chain lengths for stationary phases provides a useful degree of retention selectivity that can be tailored to protein size and structure. Despite polar side chains being found on about half of the proteinogenic amino acids, the hydrophobic areas, coupled with the large size of proteins, enables them to be well retained by reversed-phase chromatography. Proteins are typically loaded in high aqueous mobile phase and then eluted between 50% and 90% acetonitrile or methanol. This approach helps exclude salts and other small molecule additives which could otherwise interfere with the electrospray ionization process. ESI is the most common ionization techniques used for establishing the molecular weight of specific intact proteins when MS is coupled with chromatography. The mass determined is typically a nominal, isotopically averaged mass unless a very-high-resolution analyser such as FTMS is used. The large number of charges that a protein typically acquires during ESI, and their wide distribution (charge state envelope), means that isotope peaks are spaced by small fractions of a dalton and are therefore usually not resolved. In these cases the monoisotopic mass is not determined, and an average mass is 'deconvoluted' from the protein charge state envelope. As well as confirmation of the intact mass of proteins, LC-MS can also be used to identify post-translational modifications (PTMs); for example, hydroxylation of specific amino acid residues. PTMs are often involved with protein signalling *in vivo*. See also Chapter 8 for a description of top-down proteomics.

7.4.6 **Metabolite profiling by LC-MS**

Metabolite analysis from biological samples can provide a particular challenge for LC-MS. Extracts from biological samples such as cells, tissues, or biofluids can contain many thousands of compounds exhibiting a range of polarities, molecular weights, and abundances. Their retention by LC prior to MS is essential to reduce ion suppression and provide a means to quantify each metabolite. The analysis of a wide range of 'small biomolecules' from natural systems often requires reversed-phase chromatography for low and medium polarity compounds and a more polar retentive chromatographic approach for polar and ionic metabolites. Various approaches, including HILIC, ion-pairing and ion-exchange chromatography (with ion-suppression technology), can be used (see Table 7.3). Multi-platform LC-MS is often not necessary when targeting a small number of compounds, but global profiling approaches, such as for metabolomics (see Chapter 8), often use multiple LC techniques to ensure broad compound coverage in biological extracts.

7.4.7 **Oligonucleotide analysis by LC-MS**

The analysis of synthetic and endogenous oligonucleotides has become increasingly important in recent years for developing new therapeutics and chemical probes. Oligonucleotides are highly polar oligomers with multiple phosphate groups producing strongly negative deprotonated molecules using ESI. Their chromatographic separation is generally challenging, with ion-pairing and mixed-mode chromatographic approaches the most common. The challenge is to be able to distinguish small changes in structure (a single change in base sequence), remove contaminating cations (metal ions such as potassium and sodium, which can form clusters with oligonucleotides that compromise sensitivity), and in some cases identify modifications such as methylation, which can be associated with gene expression and epigenetic regulation in cells. Where double-stranded oligonucleotides are analysed, it is advantageous to operate the LC column at temperatures above the melting temperature (T_m) of the dimer so that the two strands are separated and detected individually. This requires the use of stationary phases with high thermal stability.

7.5 **Gas chromatography mass spectrometry (GC-MS)**

Gas chromatography was first coupled with mass spectrometry (GC-MS) in the 1960s. The mechanism of separation is based on differential partitioning of analytes between a mobile phase and a stationary phase (as for LC), but in GC the mobile phase is a gas (usually helium, but sometimes nitrogen or hydrogen). There are three main types of interaction that take place leading to separation of analytes: dispersion interactions (van der Waals), permanent dipole–dipole interactions, and hydrogen bonding (where applicable). All three can play a role in the retention of molecules on a GC column.

Modern GC-MS is performed using capillary columns that have polysiloxanes bonded to the inside surface, and these tend to have high separation efficiencies.

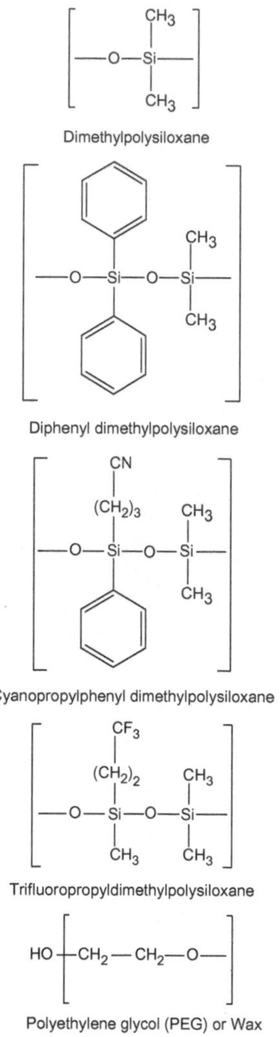

Dimethylpolysiloxane

Diphenyl dimethylpolysiloxane

Cyanopropylphenyl dimethylpolysiloxane

Trifluoropropyldimethylpolysiloxane

HO—[CH₂—CH₂—O]—

Polyethylene glycol (PEG) or Wax

Figure 7.6 Monomer structures of polymeric groups bonded to the inside of capillary columns to enable retention of compounds by GC-MS. Different proportions of these polymers are used to provide specific retention capabilities.

Table 7.5 Relative polarity of polymeric groups used for GC-MS stationary phases.

GC stationary phase	Polarity
Dimethylpolysiloxane	+
Diphenyl dimethylpolysiloxane	++
Cyanopropylphenyl dimethylpolysiloxane	+++
Trifluoropropyldimethylpolysiloxane	++++
Polyethylene glycol (PEG) or Wax	+++++

Retention on a GC column works on the principle that stationary phase functional groups are of a similar polarity to the analytes they retain. By definition, gas-phase analytes tend to be relatively small and non-polar molecules, and this is reflected in most column chemistries of which there are five basic types. Figure 7.6 shows their structures that represent a range of polarities (see Table 7.5). A more polar column will generally be chosen to separate more polar molecules. Manufacturers mix these chemistries in different proportions to provide a particular column polarity, allowing the design of a very wide range of stationary phase types.

Analyte molecules are partitioned between the gaseous mobile phase, which carries them through the column, and the stationary phase, leading to their retention (Figure 7.7). A temperature gradient is commonly used to elute compounds, and their elution time is therefore usually based on volatility.

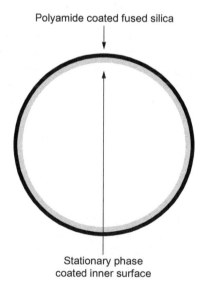

Polyamide coated fused silica

Stationary phase coated inner surface

Figure 7.7 Cross-section of a GC capillary column.

As electron ionization (EI) requires analytes to be vaporized in the gas phase it is ideally suited for direct coupling to GC. Chemical ionization (CI) is similarly straightforward to couple, and today EI and CI ion sources are still the most common ionization techniques used for GC-MS.

7.5.1 GC-MS performance and applications

The differences between gas-phase and liquid-phase analyte interactions with a stationary phase lead to some characteristic differences between GC-MS and LC-MS. For example, gas-phase kinetics are inherently faster than liquid-phase kinetics, providing greater efficiency and often faster elution and narrower peaks in GC. A narrower peak means it will be higher in intensity for the same concentration of analyte (peak area must remain constant because it is proportional to analyte concentration). GC-MS can therefore be more sensitive than LC-MS when it is possible to measure the same analyte, but there does not tend to be much direct overlap, due to their differing physical requirements. The need for gas-phase molecules places a significant restriction on the types of analyte that can be measured by GC. Also the relatively wide range of stationary phases available for LC are not matched in GC. This is mostly because polar and ionic species do not generally meet the volatility requirement of GC separation systems and are therefore not directly suitable as analytes. It is possible to analyse more polar, hydrophilic compounds by GC-MS, but they first require derivatization to increase their volatility. This is the process of chemically modifying functional groups to make the whole analyte more volatile and/or more stable to the elevated temperatures associated with GC analysis. This approach can work well, although careful attention is often needed to choose the right derivative, which usually requires prior knowledge of sample composition (Table 7.6 and Figure 7.8).

The narrow width of rapidly-eluting GC peaks requires fast scan times by mass spectrometry to ensure adequate ion abundance measurements across an

Table 7.6 A list of common derivatizing agents used for GC-MS.

Derivative type	Examples	Typical compounds
Silylation	BSTFA, BSA, TMSI, TMSI, TMCS, BSTFA, HMDS	Acids, alcohols, thiols, amines, amides, some ketones and aldehydes
Acylation	Acyl halides and fluorinated anhydrides including PFPA, HFBA, TFAA. N-Acetylimidazole	Alcohols, phenols, carbohydrates, amides, amines, carbonyl, hydroxyls, sulfonamides, amino acids
Alkylation	PFBBr, TMAH, m-TFPTAH MCF, diazomethane	Organic acids, alcohols, phenols, sulphonamides, steroids, oligosaccharides

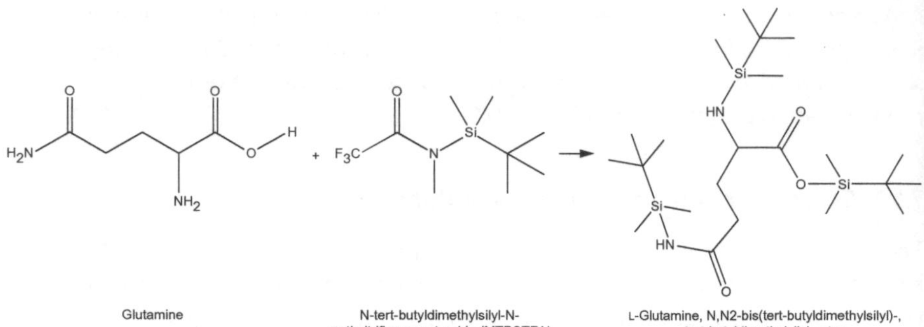

Glutamine   N-tert-butyldimethylsilyl-N-   L-Glutamine, N,N2-bis(tert-butyldimethylsilyl)-,
               methyltrifluoroacetamide (MTBSTFA)   tert-butyldimethylsilyl ester

Figure 7.8 The derivatization of glutamine using N-tert-butyldimethylsilyl-N-methyltrifluoroacetamide (MTBSTFA). This reaction leads to silylation formation of the glutamine tert-butyl dimethylsilyl derivative. In general, MTBSTFA reacts with active hydrogen containing polar side groups. In the case of glutamine the two amine groups and the carboxylic acid are derivatized, making the whole molecule less polar and more volatile.

eluting peak. GC-MS therefore benefits particularly from fast scanning analysers and acquisition modes which optimize scan speed. TOF and quadrupole mass analysers are typically used in GC-MS experiments, and SIM, SRM, and MRM acquisition modes are commonly employed to maximize sensitivity. Recently, GC-orbitrap instruments have been developed commercially. These offer very high resolution and mass accuracy stability with the benefits of GC separation. GC-MS provides one of the most sensitive and quantitative tools for the analysis of individual compounds in complex mixtures. It is used for environmental monitoring where absolute levels of trace toxins or environmental pollutants are required. It is also employed for gas analysis and the measurement of volatile compounds associated with synthetic and biological chemical systems. See Chapter 8 for further applications.

7.6 Ion mobility separations and mass spectrometry

Ion mobility spectrometry (IMS) is a gas-phase ion separation technique that resolves molecules based on their charge and cross-sectional area. It is a method of separation which can be used in conjunction with mass spectrometry in a hybrid configuration known as IMS-MS.

During the IMS process, ions experience an electric field which propels them though a cell or tube within which an inert gas is present. The resistance that the ions experience due to the presence of gas molecules is a function of their cross-sectional area, and hence their transit time is determined by their orientation-averaged cross-sectional profile. Their 'drift time' through the cell can be used to differentiate molecular species that have different cross-sectional areas but the same m/z (Figure 7.9).

Ion mobility techniques enable separations on the millisecond timescale. Although many IMS systems are used as stand-alone analysers for the detection of explosives, pharmaceuticals, and toxins, for example, their coupling with high-resolution mass spectrometers provides an additional and orthogonal separation of analytes. TOF analysers are particularly suitable for coupling due to their

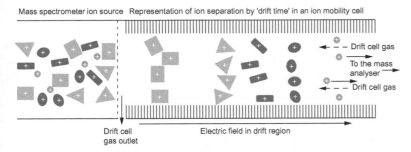

Mass spectrometer ion source   Representation of ion separation by 'drift time' in an ion mobility cell

Figure 7.9 A schematic illustrating a drift cell incorporated into a mass spectrometer system just after the ion source. Ions formed in the source are separated based on their differential transit time through the drift cell. There are a number of different types of IMS devices now incorporated into mass spectrometers. These include travelling wave IMS, trapped IMS (TIMS), field asymmetric IMS (FAIMS), and drift tube IMS (DT-IMS). In some cases their location within the mass spectrometer can also vary. In this schematic the drift cell is located just after the ion source region, but for travelling wave IMS, for example, other configurations include locating the drift cell after a quadrupole or ion trap. This format enables drift-time separation of product ions in a tandem mass spectrometry experiment.

fast scan times, providing the resolution to capture small differences in drift time. A number of hybrid tandem and non-tandem MS systems are available commercially with IMS cells now incorporated into the design.

Applications of IMS-MS include small molecule analysis such as pharmaceutical characterization and increasingly metabolomics, but many applications of IMS-MS to date have focused on the analysis of protein conformation in the gas phase. Figure 7.10 shows a two-dimensional IMS-MS drift-time versus m/z

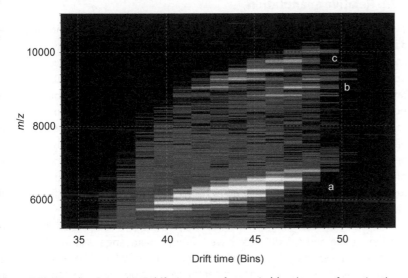

Figure 7.10 Two-dimensional IMS drift-time map of a protein (a) and two conformational states of its dimer (b and c). Colour intensity represents ion abundance. Although the dimer charge states have the same m/z they are separated from each other by drift time in the IMS cell due to their differing conformations and hence shape profile.

plot associated with a protein ionized by ESI. Two conformational states of the protein dimer can be clearly seen in the two-dimensional drift-time vs m/z plot. They have the same m/z value but a different drift time representing differences in their three-dimensional shape. Conventional mass spectrometry would not be able to differentiate these two structures, but they nevertheless may have quite different biological functions *in vivo*.

7.7 Quantitation by mass spectrometry

The aim of a quantitative mass spectrometry experiment is to determine the m/z value of an analyte of interest and use its abundance to calculate the analyte's concentration. There are two basic approaches to this, referred to as 'relative' and 'absolute' quantitation. Each provides different information about the amount of analyte detected. Relative quantitation evaluates the relative abundance of a compound in two or more samples by comparing the ion abundance of representative m/z values. It assumes a linear relationship between concentration and the number of ions of the compound. This is useful when the purpose of an experiment is to know whether the concentration of a particular analyte changes from one sample to the next, and in which direction and the magnitude of the change. However, it does not provide an absolute amount of analyte in concentration units. This is what is provided by absolute quantitation, which determines the actual concentration of the analyte in appropriate concentration units. Which approach is most useful depends on the aims of the experiment.

7.7.1 Relative quantitation

The combination of a chromatograph and a mass spectrometer provides the ability to isolate compounds from a mixture and make an abundance measurement. Both chromatographic peak height and peak area are proportional to analyte concentration. However, care needs to be taken because relative quantitation often assumes a linear detector response with concentration but this may not always be the case for two main reasons. The first is that the sample matrix may differ between samples and interfere to differing extents with the ionization efficiency of the same analyte. As we have seen, this can potentially affect the response in one sample relative to another (Section 7.2.1). The chromatographic process by definition reduces the sample complexity at any particular point during the elution process, but if samples have quite different matrices this can still have a significant effect on relative peak areas.

The second reason why a non-linear response may be obtained is because the ion source and MS detector have a finite 'linear range' within which their response is proportional to concentration outside of which the relationship is no longer valid. All types of mass detector have a non-linear response to analyte concentration outside their physical processing limitations which are found above and below certain analyte concentrations.

'Sample matrix' is the molecular composition of a sample other than the specific analytes of interest. This is a common terminology in LC-MS because it can have an impact on the ionization efficiency of an analyte and therefore affect sensitivity and quality of quantitative results.

7.7.2 Absolute quantitation

An absolute quantitation experiment enables some of the biases and non-linear aspects of an LC-MS experiment to be evaluated and minimized, as well as providing a calculation of the absolute analyte concentration. There are three different strategies to achieve this: (i) using an external calibration, (ii) creating an internal calibration by isotope dilution, and (iii) creating an internal calibration by standard addition. Each approach has its own particular benefits and challenges. Ultimately, which method is chosen depends on the accuracy of the quantification required and the nature and quantity of the sample available. All the approaches to absolute quantification are based on using authentic standards of known concentration to predict the concentration of the analyte in a sample.

7.7.3 External calibration

The external calibration method for absolute quantification is perhaps the simplest approach and most universally applicable. The ion intensities or peak area responses for a series of different concentrations of the authentic standard are measured, and then a plot of concentration versus peak area is created. This represents the relationship between analyte concentration and detector response. Figure 7.11 shows the experimental relationship between peak area and concentration from the analysis of a pharmaceutical drug called Zolmitriptan, using LC-MS (UHPLC connected to a Thermo Q-exactive orbitrap).

If a straight line can be drawn through the points linking peak area response and concentration it demonstrates a linear relationship. In practice, least-squares linear regression is used to calculate the equation of the line of best fit through the data points, and the efficiency of the fitting is determined by how close the R^2 value is to 1. The equation for the line (which is usually calculated automatically

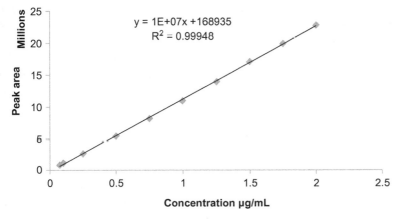

Figure 7.11 Shows the concentration and peak area relationship determined by a C18 reversed-phase LC-MS analysis of Zolmitriptan, a selective serotonin receptor agonist used in the treatment of migraine.

in many graph plotting programs) can then be used to determine the unknown concentration of an analyte in a sample from its peak area measurement.

When performing an external calibration, standards should mimic the samples as accurately as possible. Often, calibration solutions are prepared in solvent and matrix-matched solvents. If the concentration of a drug is being quantified in urine, for example, then the calibration standard should ideally be made up in urine (which does not contain the analyte). This is especially important when complex matrices are analysed but not always possible; for example, when the analyte is endogenous to the matrix such as for endogenous metabolites in blood plasma. In some cases a pure solvent is the best match that can be achieved, but in this case it should be remembered that matrix effects could affect the accuracy of the quantification.

The linear part of the calibration curve should extend beyond the concentration of the analyte in experimental samples. This may seem obvious but the linear range is finite for any calibration curve although the equation generated is not. It is not appropriate to use the equation for the line if the analyte response falls outside the data points used to construct the calibration. Concentrations of the standard should also ideally be equally spaced and at least six different concentrations used. Finally, it is good practice to make the calibration standards a number of times and run each multiple times to ensure reproducibility.

7.7.4 Internal standards

All sample preparation procedures and analytical measurements are subject to error, and it is useful to be able to quantify these errors and find strategies to minimize their effects. A natural variability in blood glucose levels between patients, for example, reflects biological differences that may be of interest, but the steps involved in preparing each of the samples and the analytical measurements themselves add experimental and analytical errors to each measurement. One of the most powerful approaches for identifying and reducing sample preparation and analytical error is to use an internal standard. This is added to samples before they are prepared for analysis. The principle is that if a known concentration of internal standard is added to each sample as the first step in sample preparation, its measurement alongside the analyte(s) allows any variability that results from sample preparation to be determined. Internal standards can be used to modulate external calibrations or as part of an internal calibration strategy (see Section 7.7.6).

7.7.5 Isotope dilution mass spectrometry (IDMS)

In addition to monitoring and correcting for variability due to sample preparation and analysis, an internal standard can also be used for quantitation. The ratio between the intensity of a suitable internal standard and the analyte can be directly proportional to concentration. For this to be the case the internal standard should, ideally, have identical physical and chemical properties to the analyte. This will then give it an identical chromatographic retention time and

An example of the calculations involved in determining the concentration of Zolmitriptan using the calibration curve information in Figure 7.11. If the analysis of a sample containing an unknown concentration of Zolmitriptan provides an experimental peak area of 13,200,101, what concentration does this represent? The equation for the line was calculated as $Y = 1e^7 * X + 168{,}935$, which is equivalent to $Y = 10{,}000{,}000 * X + 168{,}935$. If Y is measured to be 13,200,101, then rearranging the equation can solve for X (the unknown concentration). $X = (13{,}200{,}101 - 168{,}935)/10{,}000{,}000 = 1.303$ µg/mL.

hence it will experience the same sample matrix effects. However, it must also have a different m/z value to distinguish it from the analyte. A different stable isotope version of the analyte meets these requirements and should be fully enriched, ideally using ^{13}C or ^{15}N (heavier isotopes), in relevant positions on the molecule.

In order to distinguish an isotopically enriched standard from a small molecule analyte, the isotope enrichment should lead to at least a 3 Da change in mass. The isotope-enriched internal standard should be added to the sample as early as possible in the process, and at a concentration close to the analyte concentration. The ratio of the signal for the internal standard and the analyte are then determined and used to calculate the unknown concentration of the analyte.

The advantages of isotope dilution mass spectrometry for quantification is that it ensures perfect matching of the sample matrix, and all measurements are under the same sample conditions (not the case for external standard calibration methods). A disadvantage of this approach is that it can lead to dilution of the sample (although this is not necessary if sample preparation involves dry-down of the sample and reconstitution in a different solvent). It also requires isotope-enriched standards to be used which are generally expensive and only commercially available for a relatively small number of compounds. Despite these drawbacks, IDMS is generally considered the most accurate and precise calibration method available for quantitation by mass spectrometry.

7.7.6 Internal calibration by standard addition

In certain circumstances, when isotope-enriched standards are not available and there is sufficient sample, an internal calibration can be made using the addition of authentic standards. In this approach, the sample is divided into four or more equal volumes, and different measured volumes of an authentic standard solution are then added to three of the four samples. The volumes of all four samples are then made up to be equal using the authentic standard diluent, and each sample is analysed by LC-MS, GC-MS or another suitable analytical method. A calibration curve is constructed in a similar way as for an external calibration, and the concentration of the original sample (the one that was not spiked with authentic standard) can be determined from the equation of the line. The advantage of this approach is that it also perfectly matches the matrix and ensures no analytical mismatch between the calibrant and the analyte. However, it uses significantly more sample than other methods; at least four runs are required, and it is less easy to use for low-abundance analytes. It can also suffer from relatively high measurement uncertainties.

The advantages and disadvantages of the various methods of quantification discussed are summarized in Figure 7.12.

7.7.7 Lower limit of detection and quantification (LLOD and LLOQ)

In general, the lower limit of detection (LLOD) for a mass spectrometry experiment refers to the lowest concentration of an analyte which can be detected

Deuterated internal standards for isotope dilution mass spectrometry are not ideal because if a large number of deuterium atoms are present in the standard it can lead to differences in chromatographic retention time between the analyte and standard, particularly when using GC-MS. This means that any matrix effects present will be slightly different for the analyte and the internal standard which could compromise effectiveness of the quantitation. ^{13}C and ^{15}N isotopically enriched forms of an analyte are generally not resolved by conventional gas or liquid chromatography, and hence the internal standard and analyte can be identified by having the same retention time but different m/z values.

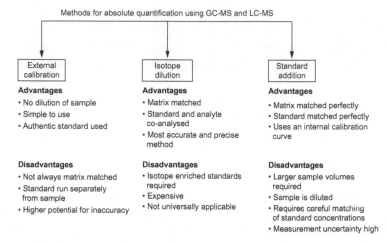

Figure 7.12 Comparison of different methods for absolute quantification by LC-MS and GC-MS.

reliably. The lower limit of quantification (LLOQ) is higher than the LLOD and refers to the lowest concentration of an analyte for which a reliable relationship between concentration and signal response can be established. Figure 7.13 demonstrates that in general the higher the sensitivity of the analysis the lower the limit of detection and smaller the measurement uncertainty and hence limit of quantification.

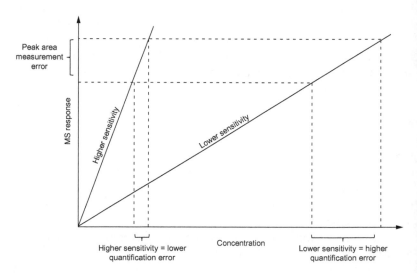

Figure 7.13 Higher sensitivity leads to an increased ability to distinguish small changes in concentration and a lower limit of detection (LLOD), as illustrated by the graph. The two separate slopes represent two different sensitivity measurements. For the same measurement error, different sensitivities lead to different levels of uncertainty in the error associated with the quantification.

In practice there are two main approaches for determining limits of detection and quantification, and which of them is used is often determined by the type of application and availability of the sample. The first approach is based on measuring signal-to-background noise ratio (S:N). In this context, LLOD is defined as the lowest concentration of an analyte which produces a signal-to-noise ratio of 3:1. The equivalent LLOQ definition is the lowest concentration of an analyte which has a signal-to-noise ratio of 10:1. Note that these definitions do not include how the background noise is to be determined, (for example, how wide should the *m/z* range chosen for the noise measurement be? How many scans should be summed?). An alternative and more rigorous approach is to determine an 'instrument detection limit' (IDL) or 'method detection limit' (MDL). These approaches aim to bring greater statistical rigour to the determination of detection limits. They have the benefit of being based on a series of analyses rather than a single analysis, but they require an authentic standard and analyte-free sample matrix, which are not always available. An explanation of how IDL and MDL experiments are performed is beyond the scope of this book, but they are well-established methods, and details can be found elsewhere (see Further reading at the end of this chapter).

7.8 Summary

From the material presented in this chapter you should now be familiar with the basic concepts associated with combining separation techniques with mass spectrometry and the particular benefits that these can provide for the analysis of complex samples. You should be familiar with ion mobility and how it separates compounds and can be combined with MS detection in various configurations. You should know how chromatographic systems enable separation of complex mixtures, and how, when coupled to mass spectrometry, these can provide unique capabilities for quantification of analyte concentrations. You should also now have an understanding of the principles behind quantitative analysis and the different strategies used to enable relative and absolute quantification using hyphenated techniques.

You should be able to explain:

- Why mass spectrometry alone is not a universally quantitative technique.
- The challenges and benefits associated with LC-MS.
- The challenges and benefits associated with GC-MS.
- The basic principles behind ion mobility coupled to mass spectrometry.
- The principles behind quantitative analysis using chromatography coupled to mass spectrometry.
- A strategy used for calculating limits of detection and quantification.

7.9 **Exercises**

7.1   Explain briefly why mass spectrometers typically do not provide a universal response to analyte concentration.

7.2   Explain the term 'ion suppression'.

7.3   Why is reversed-phase chromatography particularly suitable for coupling to ESI-MS?

7.4   Explain why the peak area of a mass extracted chromatogram is directly proportional to analyte concentration.

7.5   Why is the scan speed of a mass spectrometer important in LC-MS and GC-MS experiments?

7.6   Why is GC-MS generally more sensitive than LC-MS but less versatile?

7.7   Name two advantages and two disadvantages of nano-LC compared to conventional HPLC systems.

7.8   How are compounds separated from one another in ion-mobility spectrometry?

7.9   What is the difference between relative and absolute quantification?

7.10 Briefly explain the advantages and disadvantages associated with calculating lower limit of detection (LLOD) by comparing the signal-to-noise ratio of an analyte.

7.11 Analysis of blood plasma produced an experimental extracted ion chromatogram (EIC) peak area of 9301 for phenylalanine. An authentic standard of phenylalanine was used to construct an external calibration curve, and $y = 200x + 34$ was the experimentally determined equation for the line. Use this information to estimate the concentration of phenylalanine in the blood plasma sample with concentration units in µmol/L.

7.10 **Further reading**

Boyd, R. K., Basic, C., and Betham, R. A. (2008). *Trace Quantitative Analysis*. Chichester: Wiley.

Lavagnini, I., Magno, F., Seraglia, R., and Traldi, P. (2006). *Quantitative Applications of Mass Spectrometry*. Chichester: Wiley.

Watson, J. T. and Sparkman, O. D. (2007). *Introduction to Mass Spectrometry*, 4th edn. Chichester: Wiley.

8 Mass spectrometry applications

8.1 Introduction

The ability of mass spectrometry to identify and quantify individual compounds from complex samples with high sensitivity has led to a wide range of physical, chemical, biological, medical, and industrial applications. In this chapter we will address selected examples, some of which are well established, whilst others are still in development. All, however, have benefited from technical advances in sensitivity, selectivity, and the ability to process mass spectrometric data more rapidly and at larger scale. Developments have enabled identification and quantification of proteins and metabolites in biological samples (see proteomics (Section 8.4.1) and metabolomics (Section 8.3.3)), and more recently, help guide surgical procedures with mass spectrometric feedback in real time. This chapter does not seek to provide an exhaustive range of applications—which would be beyond the scope of the book—instead we have selected examples that serve to demonstrate the breadth of the field and the enabling capability of new technological developments. The first part of the chapter deals with small-molecule applications, in particular related to the environment, sports, biology, and medicine, and the second part focuses on the characterization of larger molecules with an emphasis on the analysis of proteins and mass spectrometry imaging.

8.2 Mass spectrometry applications in the environment and sport

8.2.1 Environmental pollutants

The last three decades of the twentieth century saw a wide range of high-profile environmental disasters. These included large oil spills (Exxon Valdez, 1989), nuclear contamination (Chernobyl, 1986), dumping of toxic industrial waste (Love Canal, discovered in 1977), contamination of drinking water, accumulation of heavy metals in sea life, and release of a wide range of air pollutants that contributed to 'acid rain', chlorofluorocarbons affecting the ozone layer, and methyl isocyanate contamination of the environment in Bhopal, India (1984).

The latter has been considered the world's worst industrial disaster of the twentieth century. These, and other environmental incidents, led to an increased public awareness of the environmental threat posed by industrial activities. Such events resulted in new legislation in many countries with a greater emphasis on closer monitoring of the environment and industrial outputs. As a consequence of this process, mass spectrometry became a key analytical tool for monitoring and in searching for evidence of harmful environmental contamination associated with industrial activities.

The high sensitivity, selectivity, and ability to differentiate multiple compounds in a single sample gave mass spectrometry a significant advantage over other analytical approaches. GC-MS, in particular, quickly became one of the most important tools for monitoring small-molecule pollutants that were often volatile and from complex chemical sources. Sample complexity and volatility lent itself well to GC separation and quadrupoles became popular mass analysers to couple to GC with their wide dynamic range and ability to perform accurate quantification.

Many volatile organic compounds (VOCs), resulting from a range of human activities, including industrial processes and motor vehicle emissions, are known to impact negatively on ozone levels, climate change, and human health. For example, a particularly pernicious group of carcinogenic VOCs, dangerous to human health, are the polychlorinated dibenzo-para-dioxins (PCDDs) and polychlorinated dibenzofurans (PCDFs), collectively known as 'dioxins'. These are released via industrial processes including bleaching of pulp in paper manufacture and incineration systems, but can also result from volcanic eruptions and forest fires. These compounds persist for long periods in the environment and accumulate through food chains, residing in the fatty tissues of humans and animals. Dioxin emissions and environmental levels are now tightly regulated, and routine analysis of air samples, in many parts of the world, is performed using gas chromatography coupled to tandem mass spectrometry.

Many VOCs can be toxic even at the parts per billion level, so it is important that samples can be concentrated for analysis. This is usually achieved by thermal desorption (TD) mass spectrometry directly coupled to a GC interface, which can continuously sample air for a period of time, trap VOCs on a stationary phase, and then evaporate these directly into the GC inlet of a GC-MS system. Figure 8.1 provides a schematic of a TD interface for sampling VOCs such as dioxins in air. A TD autosampler is coupled directly to a GC-inlet that is connected directly to a triple quadrupole MS system (TD-GC-MS/MS). SRM for a specific VOC is often used to perform simultaneous identification and quantification. The SRM transition provides selectivity, whilst the MS/MS product ions are used to quantify with enhanced signal-to-noise (compared to the precursor).

A relatively recent technical development to benefit food and environmental analysis is the GC-orbitrap. Introduced commercially in 2015, this hyphenated instrumentation couples a GC (EI/CI) to an orbitrap mass spectrometer (Thermo

Dioxins are a class of toxic and highly persistent pollutants produced from industrial processes and waste incineration. Their levels are carefully monitored using GC-MS/MS methodology.

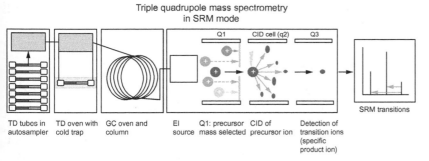

Triple quadrupole mass spectrometry
in SRM mode

| TD tubes in autosampler | TD oven with cold trap | GC oven and column | | EI source | Q1: precursor mass selected | CID of precursor ion | Detection of transition ions (specific product ion) |

Figure 8.1 Analysis of volatile organic compounds (VOCs) using TD-GC-MS/MS.

Electron, MA, USA). It provides higher resolution and mass accuracy stability than other types of mass analyser and providing potential for gaining greater insights into complex samples than for quadrupole or time-of-flight instruments for example. These and other high-resolution analysers are starting to replace quadrupole detectors in environmental monitoring, and this trend is likely to continue.

8.2.2 **Radiocarbon dating**

Modern radiocarbon dating uses accelerator mass spectrometers to measure the ^{14}C concentration of a sample which was once part of a living organism in equilibrium with atmospheric CO_2. With a half-life of 5730 years, the amount of radioactive ^{14}C still present is proportional to the 'age' of the sample after the death of the organism. This can be used to calculate the 'age' of material up to a practical limit of approximately 50,000–60,000 years. This technique has afforded an almost unique ability to create accurate chronologies from the analysis of plant, animal, and human remains. As a consequence this has led to a much clearer picture of prehistoric cultural evolution, the spread of agriculture, and the migration of modern humans, than would otherwise have been be possible.

^{14}C is produced in the stratosphere on a continuous basis via collisions between cosmic ray-produced neutrons and ^{14}N atoms (see equation 8.1). About 8 kg of ^{14}C is formed per year. This mostly reacts with oxygen to form $^{14}CO_2$, which mixes rapidly within the atmosphere. Photosynthetic mechanisms subsequently sequester $^{14}CO_2$ alongside $^{12}CO_2$ into plants. The ^{14}C then finds its way into virtually all living organisms via interconnected food webs. Because almost all living tissues are continuously degraded and reformed (in a process known as 'turnover'), ^{14}C abundance in an organism stays constant and at equilibrium with atmospheric $^{14}CO_2$ levels whilst an organism is alive. After death the ^{14}C levels (no longer at equilibrium with atmospheric CO_2) begin to reduce according to the ^{14}C radioactive decay half-life of 5730 years (see equation 8.2).

Accurate measurement of the ^{14}C content of once-living materials therefore provides the opportunity to determine the amount of time that has elapsed since

The 'half-life' of a radioactive isotope is the time it takes for one-half of atoms present in a sample to be transformed via radioactive decay.

$$n + {}^{14}_{7}\text{N} \rightarrow {}^{14}_{6}\text{C} + \text{p} \qquad (8.1)$$

The number of ^{14}C atoms present in the atmosphere fluctuates to a small extent (associated mostly with cosmic ray flux) but is at a ratio of roughly around one ^{14}C atom per trillion ^{12}C atoms (10^{-12}). In a sample of organic material 37,000 years old (since time of death), the ^{14}C content will have diminished to around one ^{14}C atom per 100 trillion ^{12}C atoms (10^{-14}). This puts into perspective the challenge faced by accelerator mass spectrometry, which must accurately discriminate ^{12}C, ^{13}C, ^{14}C, and ^{14}N produced from organic samples with at least a one-in-a-billion abundance ratio.

$$^{14}_{6}C \rightarrow\ ^{14}_{7}N + e^{-} + \bar{\nu}_e \qquad (8.2)$$

death. Accelerator mass spectrometry has emerged as the most efficient and sensitive way to make accurate ^{14}C measurements.

8.2.2.1 Sample preparation

A wide range of materials including wood, bone, food remains, charcoal, cloth, CO_2 in ice cores, and even whisky are radiocarbon dated, but a date can only be considered valid if the carbon component is wholly derived from the material to be dated. That is to say, contaminating carbon sources can have a significant detrimental effect on the accuracy of ^{14}C measurements and hence interpretation of the age of a sample. A great deal of time, effort, and research goes into removing carbon-containing contamination. To do this, samples may be treated by sequential washing with dilute HCl (removes carbonates and oxalates), followed by dilute NaOH to remove humic acids (derived from breakdown of plant materials in soils), followed by re-acidification. Some materials, such as bone or insect remains, are chemically better defined by isolation of the protein collagen, or the organic polymer chitin for radiocarbon dating.

Prior to analysis by accelerator mass spectrometry, collagen samples, for example, are converted into solid carbon, through initial combustion to CO_2 (usually in a continuous flow elemental analyser) followed by reduction to form graphite (often using iron). A 1 mg graphite sample is often sufficient for AMS analysis and determination of a radiocarbon date.

With the goal of providing ever better-defined materials for dating, single compounds have also been isolated and purified. In the dating of bone, for example, extraction of collagen protein is sometimes followed by digestion and purification of the individual amino acid hydroxyproline (which is unusually abundant in collagen making up approximately 10% of the amino acid residues). This provides increased specificity and the ability to simultaneously remove additional contaminating compounds through chemical isolation.

8.2.2.2 The accelerator mass spectrometer

An accelerator mass spectrometer differs in a number of ways from most of the mass spectrometers dealt with so far in this book. The instrument is not designed for mass-range coverage or indeed m/z accuracy, but rather high sensitivity and precision in discriminating isotopes of the same element. The same basic components of the mass spectrometer discussed in Chapter 1 are present, including the ion source, ion guides, mass analyser, and detector (see Chapter 1 and Figure 8.2). However, the instrumental design is also unique in several key aspects associated with ion transmission and detection. The measurement process starts with the formation of negative carbon ions from the graphite sample. These are generated through interaction with a caesium sputter ion source. The newly formed ions are then accelerated to much higher ion energies compared to most other mass spectrometer types (up to several MeV). As well as $^{14}C^{-}$ anions, other elemental and molecular anions also contribute to the total ion beam at this stage, including $^{13}CH^{-}$. A 'beam bending' magnet reminiscent of

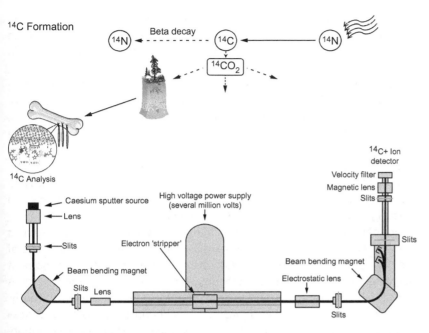

Figure 8.2 The process of ^{14}C formation in the upper atmosphere and sequestration into living matter via $^{14}CO_2$ (top). A schematic of an accelerator mass spectrometer used to measure the ^{14}C content of a sample for radiocarbon dating (below).

the magnetic sector design (Chapter 3) is used to remove any neutral molecules and isolate $^{14}C^-$. This helps remove some (but not all) $^{12}C^-$ and $^{13}C^-$ ions that predominate but are not of interest for radiocarbon dating. Next the negative ions in the ion beam are stripped of electrons by collision with (usually) oxygen or argon gas molecules, and any remaining sample molecules are atomized at this point. The process results in multiply charged positive ions (a mixture of C^+, C^{2+}, C^{3+}, and C^{4+} for each carbon isotope) that are accelerated back down to ground potential, reaching a final energy of 10 MeV or more. The next part of the instrument is reminiscent of the traditional magnetic sector mass analyser, and splits the ion beam into a number of ions of different m/z value, differentiating the remaining $^{12}C^+$ and $^{13}C^+$ ions from $^{14}C^+$ ions, which then arrive at an ion detector via a velocity filter. This is a significant achievement in selectivity, given the overwhelming number of stable carbon atoms present in any sample. Figure 8.2 displays the process of ^{14}C formation and sequestration into living matter, and a schematic of the accelerator mass spectrometer used to measure the $^{14}C^+$ content of samples.

8.2.3 Anti-doping in sports

Cultures worldwide have historically used bioactive compounds to enhance physical performance. For example, the ancient Greeks exploited a form of opium to 'enhance sporting prowess'. However, a wide range of performance-enhancing compounds are now banned in modern sporting competitions by the World Anti-Doping Agency (WADA), and a programme of testing athletes for evidence of the use of these banned compounds is now an ongoing commitment in national and international sport.

Identifying the use of banned substances is, however, a major analytical challenge, and WADA issues an updated annual 'Prohibited List' containing classes of compound that are not permitted. This includes stimulants, narcotics, hormones, anabolic compounds, beta-blockers, cannabinoids, growth factors, and masking agents. Mass spectrometry has become a critical tool for testing athlete's urine and plasma samples to screen for the use of performance-enhancing drugs. Two strategies have emerged: 1) the targeting of known prohibited compounds and their metabolic products, and 2) searching for evidence of misuse of new and emerging drugs not yet on the WADA Prohibited List.

One of the most common classes of banned substances mis-used in sport is the synthetic anabolic-androgenic steroids. These are generally precursors of the naturally occurring hormone testosterone, which promotes development of greater muscle mass and strength; but with hundreds of known derivatives (mostly alkylated forms) that break down to produce testosterone in the body, detection of mis-use is a challenging task. In addition, several of these testosterone precursors are used therapeutically and are available, via prescription, for treatment of muscle-wasting disorders.

The complexity of samples, and the potential for analytes of interest to be at very low levels, requires highly sensitive, robust, and reproducible methods. These criteria have traditionally been met by triple quadrupole mass analysers with prior inline LC or GC separation. GC-MS/MS was once the predominant technique and remains essential, but LC-MS/MS has increasingly become the primary method of choice for an increasing number of assays.

LC-MS/MS methods with triple quadrupole analysers often use precursor ion scanning coupled with UHPLC. Selected product ions from precursors of known drugs are often used for identification with high selectivity and sensitivity. Figure 8.3 provides a schematic of the generalized workflow for sample preparation and triple quadrupole precursor ion analysis used to identify banned anabolic steroids. This approach has the advantage of being fast and

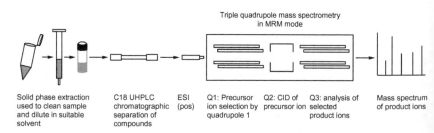

Triple quadrupole mass spectrometry
in MRM mode

| Solid phase extraction used to clean sample and dilute in suitable solvent | C18 UHPLC chromatographic separation of compounds | ESI (pos) | Q1: Precursor ion selection by quadrupole 1 | Q2: CID of precursor ion | Q3: analysis of selected product ions | Mass spectrum of product ions |

Figure 8.3 A schematic representing the analytical system for identifying banned substances using precursor ion scanning triple quadrupole mass spectrometry coupled with chromatographic separation (ESI-MS/MS).

sensitive, making it suitable for scanning a relatively large number of precursor ions in a single experiment, per sample.

Figure 8.4 shows the similarity in structure between testosterone and methyltestosterone. Testosterone is the natural endogenous steroid which has a precursor $[M+H]^+$ m/z value of 289. Methyltestosterone is used as a medical treatment for conditions which result in low testosterone. Use of methyltestosterone in sport is banned by WADA, but is used illegally by some athletes to boost testosterone levels. The chemical structure of testosterone, a significant part of many synthetic anabolic steroids misused by athletes, leads to two distinct product ions at m/z 97 and m/z 109, via low-energy collision-induced dissociation (CID) LC-MS/MS.

The fragments in Figure 8.5 are also generated by the CID fragmentation of a range of other synthetic testosterone-producing drugs, including methyltestosterone. A range of anabolic steroids can be identified by a combination of their precursor m/z values and typical 'testosterone-like' fragment ions. Figure 8.5 compares a representative product ion spectrum of testosterone and methyltestosterone. Methyltestosterone has a nominal precursor $[M+H]^+$ m/z value of 303 and testosterone has a nominal precursor $[M+H]^+$ of 289. Comparing the spectra, it can be seen that both compounds produce the same diagnostic product ions at m/z 97 and m/z 109 but can be distinguished by their different precursor m/z values. A number of lower-abundance fragments, such as m/z 159 and m/z 227, provide complementary distinguishing information.

The identification of drugs that are not yet on the WADA prohibited list is a more challenging analytical task. Rather than scanning large numbers of precursors, the ability to identity and predict chemical formulae becomes more important. This often requires higher resolution and higher mass-accuracy instruments such as Q-TOF and Q-orbitrap tandem MS systems which can provide chemical formula prediction of fragments as well as the ability to scan a wider mass range. The latter is extremely useful for understanding precursor and product ion structural relationships.

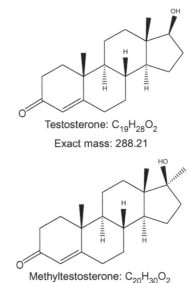

Testosterone: $C_{19}H_{28}O_2$

Exact mass: 288.21

Methyltestosterone: $C_{20}H_{30}O_2$

Exact mass: 302.22

Figure 8.4 The structures of testosterone and methyltestosterone. Testosterone is the naturally occurring steroid found endogenously and methyltestosterone is on the WADA banned list of anabolic steroids.

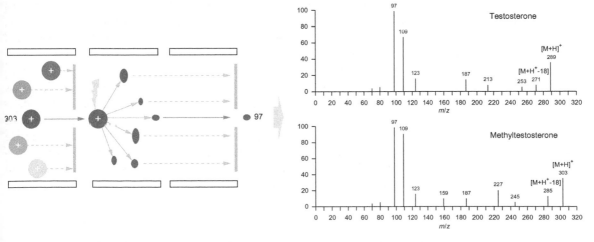

Figure 8.5 MRM analysis of synthetic testosterone derivatives in urine using precursor ion scanning.

8.3 Small-molecule mass spectrometry in biology and medicine

8.3.1 Drug discovery and development

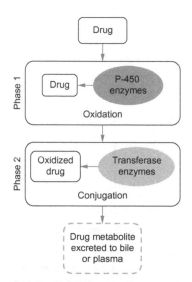

Figure 8.6 Drug metabolism in the liver: enzymes in the liver metabolize exogenous compounds such as drugs. Phase 1 generally involves oxidation via P-450 enzymes, and Phase 2 leads to conjugation of the oxidized form with other functional groups to make the compound more soluble and easily excreted.

Pharmacodynamics refers to how a drug affects the physiology and biochemistry of an organism. Pharmacokinetics focuses on how the biochemical processes of an organism affect a drug and its breakdown. Biotransformation refers to how a drug is metabolized *in vivo* and the chemical species derived from this process.

Since the 1960s, mass spectrometry has played an increasingly important role in both drug discovery and drug development. It was initially used as a tool for characterizing the molecular weight of candidate drugs, but increased selectivity, specificity, sensitivity, and an ability to interrogate data with more powerful software, has led to its application in pharmacodynamic, pharmacokinetic, and biotransformation studies. Although the actual discovery of a new drug is still most commonly accomplished by high throughput screening methods (using non-mass spectrometric approaches), it has become important to understand the structural, chemical, and biological properties of candidate drugs by using mass spectrometry.

Biotransformation studies, which focus on understanding a drug's uptake, breakdown, and metabolism, use mass spectrometry to provide valuable information about the behaviour and effectiveness of a drug *in vivo*. Potency against a molecular target by no means guarantees the success of a drug in development; indeed drug development failures are most commonly due to off-target effects that lead to toxicity. Understanding the way in which a drug interacts in a biological context is challenging, but the high financial cost associated with failure during clinical trials later in the process means it is vital to understand how a drug affects physiology and how physiology affects a drug, as early as possible in the development process. Mass spectrometry has become essential for this task.

The metabolism of a drug *in vivo*, primarily involves chemical conversion from a hydrophobic structure (which provides a drug with cell penetrating capability) to a more hydrophilic one that facilitates excretion by the organism. Enzymatic processes involved in drug metabolism are commonly oxidative in so-called 'Phase 1' metabolism, followed by 'Phase 2' attachment of more polar groups that involve chemical transformations such as glucoronidation and sulfation (see Figure 8.6). Figure 8.7 shows, as an example, the basic steps in the metabolism of aspirin in the liver.

Triple quadrupole LC-MS/MS has been used to identifying how drug candidates are metabolized *in vivo*. The flexibility of the instrumental layout means that neutral-loss scanning (see discussion in Chapter 5) can be used to identify compounds from biological samples, which contain structural elements derived from the drug candidate. In a neutral-loss scan, the first quadrupole scans all possible masses. The second quadrupole acts as a collision cell, fragmenting all arriving ions, and the third quadrupole scans at an offset from the first quadrupole by a mass difference corresponding to the neutral loss associated with the parent drug of interest (Figure 8.8). There are, however, challenges associated with the low resolution and low mass accuracy of triple quadrupole mass spectrometers that make them a poor approach for direct candidate identification.

High-resolution tandem mass spectrometers are not as useful as triple quadrupoles for neutral-loss experiments, but the increased isotope resolution of the former makes it possible to predict chemical formulae and subsequently be more confident about metabolite identifications. One approach is therefore to use data-directed high-resolution tandem mass spectrometry (DDA, see Chapter 5) in both positive and negative ion modes. Because a metabolized drug is often in a conjugated form, it commonly contains structural elements similar to the parent drug, and can therefore share some identical product ions and other common spectral characteristics. High-resolution LC-MS/MS product ions in particular are increasingly being used to identify metabolized drug forms.

Background subtraction of mass spectra is a useful way to reduce their complexity and compare control and drug-treated samples to reveal metabolites which appear only in the treated samples. Another approach is called 'mass defect filtering' (MDF) using high-resolution mass spectrometry data. The difference between the nominal mass and accurate mass for a drug metabolite is usually very close to that of the original drug, as both contain similar elemental compositions (Table 8.1). So, excluding ions that fall outside a narrow mass defect window (typically 50 mDa) is an approach which is used to narrow down potential drug metabolite candidates.

No knowledge of possible drug metabolism is needed to conduct these experiments, and they can potentially provide the mass of multiple conjugated forms,

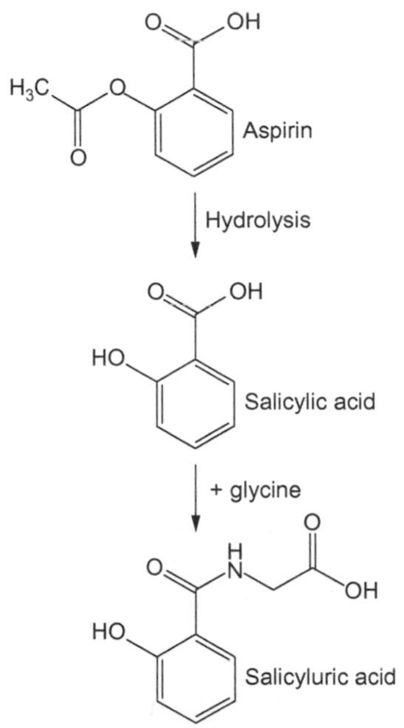

Figure 8.7 Bio-transformation of aspirin. Once ingested, aspirin enters circulation, and mostly in the liver, is hydrolysed via Phase 1 biotransformation to the active form salicylate. In Phase 2 metabolism, salicylate is then conjugated with glycine to form salicyluric acid which is excreted via plasma into the urine.

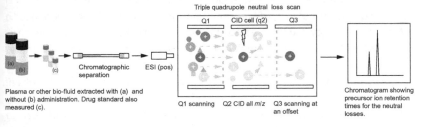

Plasma or other bio-fluid extracted with (a) and without (b) administration. Drug standard also measured (c).

Triple quadrupole neutral loss scan

Q1 scanning   Q2 CID all m/z   Q3 scanning at an offset

Chromatogram showing precursor ion retention times for the neutral losses.

Figure 8.8 Biotransformation studies using neutral loss scanning. The schematic shows how neutral loss scanning with a triple quadrupole mass analyser can be used to discover drug metabolites. Different metabolites of a drug are likely to undergo similar neutral losses when fragmented, due to their similar structures. The first quadrupole is therefore scanned across a suitable mass range. The second quadrupole fragments all scanned precursors, and the third quadrupole is offset by the neutral loss (determined by analysis of original drug or common metabolite conjugations). Compounds showing neutral losses provide a chromatogram with the retention time and intensity of the precursor. These candidates can then be subjected to further tandem mass spectrometry to determine their structure.

Table 8.1 Mass defect filter (MDF). Many biotransformation products of exogenous drugs have a mass defect which is < 50 mDa from its parent drug molecule. It can be seen that the major metabolites from the metabolism of aspirin also have an MDF < 50 mDa from aspirin. Glyceraldehyde-3-phosphate (an endogenous metabolite not related to aspirin) in contrast has an MDF significantly greater than 50 mDa (955 mDa). An MDF cut-off filter of +/– 50 mDa from high-resolution m/z data can be used to create a short-list of potential drug metabolites.

Compound	Formula	MDF (mDa)*	MDF < 50 mDa to aspirin?
Aspirin	$C_9H_8O_4$	0	Yes
Salicylate	$C_7H_6O_3$	10.5	Yes
Salicylurate	$C_9H_9NO_4$	21.1	Yes
Glyceraldehyde-5-phosphate**	$C_3H_7O_6P$	955	No

* MDF relative to aspirin.** Endogenous metabolite close in mass to aspirin.

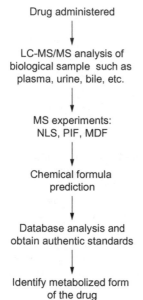

Drug administered

↓

LC-MS/MS analysis of biological sample such as plasma, urine, bile, etc.

↓

MS experiments: NLS, PIF, MDF

↓

Chemical formula prediction

↓

Database analysis and obtain authentic standards

↓

Identify metabolized form of the drug

Figure 8.9 A schematic of a typical workflow used for biotransformation studies by mass spectrometry. NLS = neutral loss scan, PIF = precursor ion fragmentation, and MDF = mass defect filtering.

A disease biomarker is a naturally occurring molecule that is characteristic of a disease, and which is used to identify the presence of that disease.

A homozygous mutation means that an identical mutation is present on both parent alleles (gene variants).

as well as multiple-product ions for each form. High m/z resolution and mass accuracy also enable chemical formula prediction of both the conjugate precursor and the product fragment ions once they are identified. A simplified schematic workflow for the determination of *in vivo* metabolic products of a drug is shown in Figure 8.9.

8.3.2 Inborn errors of metabolism

Mass spectrometry has been used in clinical practice since the early 1990s, when it became an important tool for inborn errors of metabolism (IEM) screening. IEMs are inherited genetic mutations (more than a 1000 have been identified to date) which are individually rare but collectively common and often lead to detrimental endogenous metabolism that can have severe health effects from infancy if left untreated. A number of these diseases are now readily identified by mass spectrometric analysis of dried blood spots from newborn infants.

The most common IEMs are related to defects in three areas of metabolism: amino acid metabolism, general organic acid metabolism, and fatty acid metabolism. They are usually the result of homozygous gene mutations which lead to loss of function in an enzyme that is vital to the breakdown of a specific metabolite. For example, phenylketonuria (PKU) is a well-known IEM resulting from a mutation in the gene encoding phenylalanine hydroxylase (PAH), an enzyme responsible for the metabolic conversion of phenylalanine (Phe) into tyrosine (Tyr) and is part of the phenylalanine degradation pathway in cells. Presence of the mutation leads to an endogenous build-up of phenylalanine to harmful levels (Figure 8.10), but early diagnosis and dietary therapy can lead to appropriate management and normalization of phenylalanine levels.

The mass spectrometry experiment for identification of IEMs involves detection of a pre-established metabolite biomarker at levels outside the 'normal' range. The analytical challenge lies in the chemical complexity of the sample matrix (usually blood), and the fact that individual biomarkers may be at low levels compared to other metabolites.

Early development work in the 1990s used GC-MS techniques to analyse derivatized blood and urine samples. Ester derivatives were prepared to make amino acids and acylcarnitines sufficiently volatile. Given the body of work and the proven efficacy of this approach it is still used, but the development of electrospray ionization (ESI), suitable for underivatized flow injection analysis (FIA), is now increasingly exploited due to the simplicity of sample preparation and high senstivity. Although liquid chromatography interfaced with ESI provides higher specificity and the potential for more accurate quantification compared to FIA, it is not commonly used, due to the significant amount of time it adds to each analysis and the need for a high-throughput, robust methodology. Flow injection analysis, with the use of stable isotope labelled metabolite standards, is commonly applied to enable quantitation of analytes from directly injected samples (isotope dilution quantification, see Chapter 7). This approach combines acceptable accuracy with a minimal analysis time of less than two minutes per sample. Triple quadrupole mass analysers are favoured for these applications because they can perform multiple tandem mass spectrometry experiments quickly, on the same sample, with high sensitivity and robustness. Figure 8.11 provides an overview of the analytical workflow for IEM analysis using triple quadrupole MS via FIA.

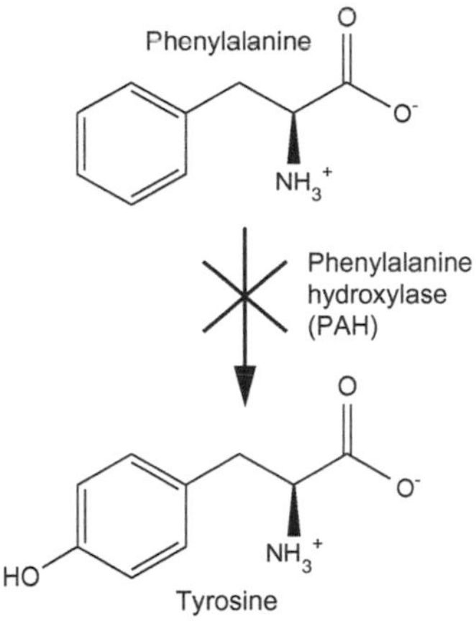

8.3.3 **Metabolomics**

Metabolomics is an experimental approach which aims to identify and quantify metabolites in complex samples in order to examine the effects of external or internal changes to a biological or environmental system. For example, it can be used to investigate the effects of diet on circulatory metabolism, study how smoking influences cardiac metabolism, search for disease biomarkers in blood plasma, determine how a gene mutation affects energy production in cancer cells, or investigate

Figure 8.10 In phenylketonuria (PKU), mutations in phenylalanine hydroxylase lead to an accumulation of phenylalanine, as it is no longer converted to tyrosine.

FIA in mass spectrometry refers to an automated method of chemical analysis in which a sample is injected into a flowing carrier solution which carries it to the ion source of the mass spectrometer directly. A liquid chromatography system can be used for this purpose by removal of the stationary phase column. This approach is most often used for automated, high-throughput delivery of samples to a mass spectrometer when separation of analytes prior to MS analysis is not a requirement.

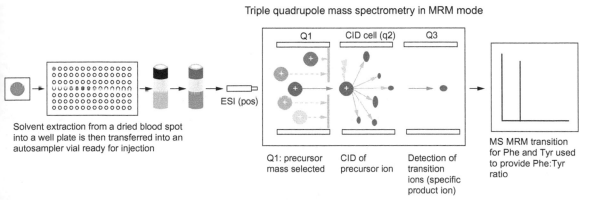

Triple quadrupole mass spectrometry in MRM mode

Solvent extraction from a dried blood spot into a well plate is then transferred into an autosampler vial ready for injection

ESI (pos)

Q1   CID cell (q2)   Q3

Q1: precursor mass selected

CID of precursor ion

Detection of transition ions (specific product ion)

MS MRM transition for Phe and Tyr used to provide Phe:Tyr ratio

Figure 8.11 A workflow showing the use of mass spectrometry for the identification of inborn errors of metabolism.

how plant pathogens perturb photosynthesis. These, and many other examples, illustrate the wide variety of potential applications of metabolomics. Statistical changes in metabolite abundance, when compared between experimental groups, are used to identify metabolites or metabolic pathways where changes can be correlated with the variables being studied. For example, 'biomarker discovery' may involve looking for metabolites that change significantly in response to a particular disease, suitable for diagnostic or prognostic purposes. Samples from a disease group of patients might be compared with those from a control group who do not have the disease. 'Pathways analysis' is an extension of biomarker discovery, but involves mapping the metabolic changes observed onto known metabolic pathways in order to understand a disease process or experimental effect at a physiologically relevant level.

Metabolomics can be used to investigate complex diseases that impact modern society, including cancer, diabetes, and dementia, which are all associated with complex genetic mutations that can have significant effects on downstream metabolism. Understanding genetic changes linked to disease helps show where potential alterations in cellular function occur and metabolomics can be used to validate these as targets for developing new drugs. Conversely, the application of untargeted metabolomics can help identify changes in metabolism resulting from as yet unidentified genetic mutations, providing information about disease processes and facilitating the development of new hypotheses.

'Wild type' refers to the normal phenotype, as opposed to a mutant form, which exhibits an altered phenotype due to one or more genetic mutations.

8.3.3.1 Analytical approaches

Estimates have suggested that the human metabolome may contain between 2500 and 8000 endogenous metabolites, and for plants this number may be over 100,000. Comprehensive analysis of samples containing large numbers of compounds provides a unique and significant analytical challenge to which mass spectrometry is particularly well suited. However, matrix effects, suitability of separation approaches, the speed of mass spectrometers, and the sensitivity and dynamic range of the mass analyser are often pushed to their technical limits by comprehensive metabolomics experiments. It is only since the early 2000s that technical capabilities have been sufficient to confidently start to meet the challenges presented.

Currently there are two basic analytical approaches taken in metabolomics: untargeted analysis and targeted analysis. An *untargeted* analysis aims to measure comprehensively *all* metabolites in a biological sample. This is technically challenging due to the wide range of chemical species (chemical structure, polarity, size, and abundance) usually present in cells, tissues, and biofluids. Currently, comprehensive coverage is the aim, but is difficult to achieve in practice; however, analysis of thousands of compounds is routinely possible, and comparison of such profiles between disease and healthy experimental groups can provide significant clues about how metabolism is affected by a particular disease, for example.

A *targeted* metabolomics analysis aims to identify and quantify a *preselected* number of metabolites. This diverges from one of the central tenets of untargeted metabolomics, that of universal coverage, but a targeted approach can be essential to increase sensitivity and confirm untargeted findings. Although comprehensive analysis is of importance, being able to identify and quantify metabolites accurately can greatly facilitate biological interpretations, and may be essential if a particular metabolic pathway is suspected of being modified in a particular experimental system. A targeted experiment places quantification and identification above comprehensive coverage, due to the inherent trade-off between sensitivity and resolution on the one hand, and maximum compound coverage, on the other.

The *metabolome* is the complete collection of small molecules, known as metabolites, found in a biological system, which might be a collection of cells, tissues, biofluids, or whole organisms.

8.3.3.2 Workflow

A metabolomics workflow incorporates: 1) extraction of metabolites from the biological system with minimal bias, 2) comprehensive measurement of relative or absolute metabolite abundances in samples, 3) identification of specific metabolites, and 4) data processing to ensure the validity of a direct comparison as well as statistical analysis and biological interpretation of the results (see Figure 8.12). One of the major differences between the application of mass

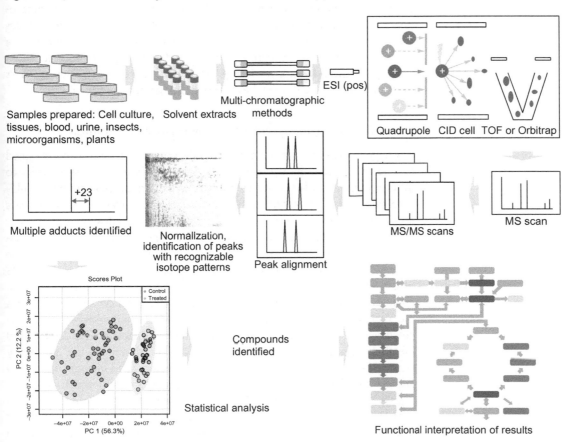

Samples prepared: Cell culture, tissues, blood, urine, insects, microorganisms, plants

Solvent extracts

Multi-chromatographic methods

ESI (pos)

Quadrupole   CID cell   TOF or Orbitrap

Multiple adducts identified

Normalization, identification of peaks with recognizable isotope patterns

Peak alignment

MS/MS scans

MS scan

Scores Plot

Compounds identified

Statistical analysis

Functional interpretation of results

Figure 8.12 A metabolomics workflow can be divided into four parts: 1) sample preparation, 2) sample analysis, 3) data processing, and 4) data interpretation.

spectrometry for metabolomics, and many other types of mass spectrometry experiments, is the aim to provide comprehensive coverage. This is fundamentally different from most other applications where the goal is to measure a specific compound or small groups of compounds. There are two basic approaches taken *within an untargeted metabolomics experiment* to try to measure as many compounds as possible. The first is to use a separation technique such as gas or liquid chromatography prior to mass spectrometry to spread out the constituent compounds in time, enabling consecutive measurements. This is advantageous because it also provides a way to remove matrix effects, limit ion suppression, and quantify compounds as well as separate them (see Chapter 7). A disadvantage is that chromatography adds a considerable amount of time to each experiment, and also introduces selectivity towards certain compounds as a result of bias from the chromatographic process.

The second approach is to use direct analysis of samples without chromatographic separation. This obviates the chromatographic selectivity bias problem and provides a molecular fingerprint without the need for prior separation. This has the distinct advantage of being fast, but it is much more challenging to provide quantitative information using this approach, particularly for comprehensive analysis. It is also more likely to suffer from matrix effects, and provides no way to distinguish between structural isomers. The optimal approach for a particular metabolomics experiment will depend on the type of information that is being sought, but currently incorporation of chromatographic separation is most commonly used despite the additional analysis time required.

8.3.3.3 Chromatography and mass analysers in metabolomics

In Chapter 7 we saw that chromatographic approaches tend to be designed around a single stationary phase-analyte interaction which has the tendency to introduce selectivity bias when complex sample matrices are analysed. A multichromatographic approach is therefore sometimes taken in metabolomics applications to maximize coverage whilst retaining the advantages that separation provides. Reversed-phase, HILIC, ion pairing, or mixed-mode chromatography, as well as derivatized and underivatized GC-MS, are often used (see Chapter 7). In contrast, a targeted approach may require only a single LC-MS or GC-MS method, and is therefore simpler and less time-consuming from an analytical perspective. The most appropriate mass spectrometer system for a metabolomics experiment is primarily determined by whether targeted or untargeted analysis is undertaken. Ideally, a versatile mass spectrometer for targeted metabolomics should be quantitative over a wide dynamic range, and highly sensitive. Although high resolution and fast analyser speed may be useful, they are less important for targeted than for untargeted experiments. Triple quadrupole MS systems coupled to GC or LC are therefore often used for targeted analysis. For untargeted metabolomics a fast scanning analyser is important, but the ability to measure accurate masses across a wide mass range is essential and makes triple quadrupoles less useful, and TOF and orbitrap analysers preferable. FT-ICR-MS systems are also used, but their relatively slow scan speeds make them currently less versatile for coupling with separation systems for the analysis of complex samples. Tandem MS methods are commonly applied (see Section 8.3.3.4) utilizing both data-directed (DDA) and data-independent (DIA) fragmentation strategies (see Chapter 5).

8.3.3.4 Identification of metabolites

The identification of large numbers of metabolites from a single sample ideally requires robust, multiple, independent measurements parameters associated with each metabolite. As the number of variables (compounds) is generally high in biological extracts, with many structural isomers present, accurate mass analysis alone is generally not sufficient for identification purposes. Indeed, identification is one of the major challenges facing metabolomics. Currently this involves incorporating accurate mass measurements, with measured fragmentation patterns, isotope abundance ratios, and chromatographic retention times. Drift-time measurements, from ion-mobility–mass spectrometry experiments, have also started to be used as an additional identification parameter and may become more common in the future (see Table 8.2).

8.3.3.5 Data analysis

An untargeted metabolomics workflow can diverge at the metabolite identification step. One approach is to pursue a statistical evaluation of all the compound measurements made between the experimental groups (irrespective of whether the compounds are identified), often using multivariate statistical analysis and visualization tools such as unsupervised (PCA) and supervised discriminant analysis (PLS-DA and OPLS-DA for example). Alternatively compounds can be first identified and then compared between experimental groups. The advantage of the second approach is that identified compounds can be mapped on to known metabolic pathways which can help with biological interpretation. The former, on the other hand, enables comprehensive evaluation of differences between experimental groups without the need to establish individually identified metabolites. These two approaches are

Principal component analysis (PCA) is a statistical method that converts a set of possibly correlated variables into a set of linearly uncorrelated variables, known as principal components. These emphasize variation and produce patterns in a dataset. By plotting principal components against each other, groupings can be identified and sample types distinguished based on their variance.

Table 8.2 Complementary measurement criteria for the identification of metabolites in metabolomics experiments.

Metabolite measurement	Distinguishing features
m/z	Determine chemical formula
Isotope pattern	Confirm elemental composition
Chromatographic retention time	Orthogonal to *m/z* and isotope pattern
Product ion fragmentation pattern	Distinguish structural differences
Collisional cross-sections (CCS) using ion mobility	Distinguishes differences in molecular shape/structure

sometimes termed 'metabolic fingerprinting' and 'metabolic profiling' respectively. In practice there are many strategies for comparing data, and the interested reader is directed to the references at the end of this chapter to explore these further. Figure 8.13a provides a principal components plot that uses all data to visualize (using a statistical model) differences and similarities between samples (and sample groups). The pathways 'heat-map' (8.13b) shows how biologically important information can be accessed when compounds are identified and mapped onto metabolic pathways.

8.3.4 Precision medicine and mass spectrometry

Techniques such as genomics, proteomics, and metabolomics have facilitated the development of a 'systems biology' approach to medicine. Since completion of the Human Genome Project it has become clearer that the genetic differences between individuals give rise to differing prevalence, severity, and response to specific diseases. Perhaps even more importantly, the susceptibility of a disease to drug treatment also varies significantly between individuals. Indeed, it has been reported that of the ten highest-grossing drugs in the USA in 2015, none could do better than an improvement in symptoms for 25% of people taking the medications (see the *Nature* article in Further Reading at the end of this chapter). This highlights an important challenge that modern medicine faces developing effective therapies: a one-size-fits-all model, which has

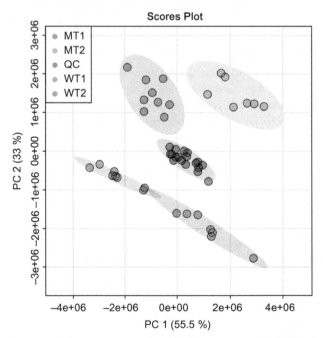

Figure 8.13a A principle components analysis (PCA) comparing mutant (MT) and wild-type (WT) cells from the same cancer cell line grown under two different concentrations of glucose ('1' and '2'). The PCA plot distinguishes the four groups, suggesting significant differences in metabolite composition between WT and MT grown on the two different concentrations of glucose. Quality control samples (QC), a mixture of all samples, lie in the centre, between experimental samples.

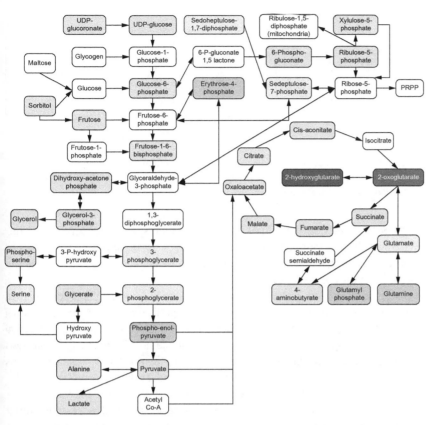

Figure 8.13b A metabolic 'heat-map' representing glycolysis, the pentose phosphate pathway and the tricarboxylic acid cycle is colour coded based on the ion intensity changes (fold-changes) between experimental groups for each metabolite. This approach can be used to visualize where changes in metabolism between experimental groups takes place.

been the *de facto* approach in biomedicine to date, appears to be significantly compromised by inherent individual differences. 'Precision medicine' or 'person-alized medicine' aims to harness understanding of these individual differences and find therapeutic responses that maximize benefit to an individual by being tailored to their biological profile.

Although it is now possible to use genomic profiling to gain insight into the genetic susceptibility of an individual to various types of illness, it has become clear that a person's genetic profile may not be sufficient to determine their personal response to drug treatment. This is likely due to the difficulties associ-ated with predicting down-stream chemistry in cells from genetic profiles alone. Additional 'omics' approaches (see metabolomics and proteomics in this chap-ter), can provide information that is closer to the chemical 'phenotype' (where genes meet the environment, for example, where a drug is metabolized). Com-bining genotype and phenotype information may be useful for predicting a person's susceptibility to a specific drug treatment. Personalized medicine, as an

'Systems biology' refers to an holistic approach to understanding biological processes and the complex systems of interaction that take place. Data gathered at a systems level, using relatively unbiased techniques such as genomics, transcriptomics, proteomics, and metabolomics, are used in computational modelling to try to understand and predict complex functional behaviour.

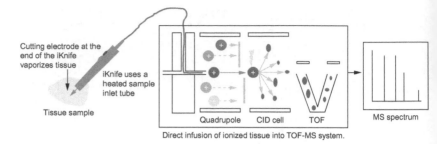

Cutting electrode at the end of the iKnife vaporizes tissue

iKnife uses a heated sample inlet tube

Tissue sample

Quadrupole   CID cell   TOF   MS spectrum

Direct infusion of ionized tissue into TOF-MS system.

Figure 8.14 A schematic of the iKnife-QTOF mass spectrometer (Waters Corporation), for which applications are being developed in a variety of medical contexts to provide greater precision in surgical procedures.

approach to treating individuals according to their specific genotype and phenotype, still has a long way to go, but mass spectrometry looks set to play a significant role in this by helping to understand how cellular processes relate to the genome in the context of disease.

Mass spectrometry is also becoming a potentially important tool for personal medicine in other ways; for example in guiding surgical procedures. A mass spectrometer system has recently been developed that is connected to a laser-based surgical blade via a sample inlet system—the so-called 'iKnife'. In this approach, MS analysis is performed on vaporized sample produced from the point at which the surgeon's blade interacts with tissue. Different tissue types produce different mass spectra. This has been designed to help guide, and make more precise, surgical treatments. For example, it has the potential to allow more precise and accurate removal of tumour tissue in surgery, providing real-time molecular information, whilst minimizing the amount of healthy tissue lost. This is an extremely important factor in many operations removing tumour tissue for example. Figure 8.14 shows a schematic of the 'iKnife' mass spectrometer system.

8.4 Analysis of proteins

8.4.1 Proteomics

As defined by the European Bioinformatics Institute, proteomics is the study of proteomes, where a proteome is a set of proteins produced by an organism, system, or biological context. Proteomic studies may be concerned with determining when and where proteins are produced, rates of protein production and degradation, protein modification, and protein interactions. Within the field of proteomics there is a need to analyse complex mixtures of proteins. Techniques are required that enable the identification and quantification of proteins with high sensitivity and precision. Mass spectrometry is able to meet these requirements, and has emerged as a vital tool in proteomic studies. Two general approaches to protein analysis have been developed: top-down and bottom-up. In the former the proteins are measured intact, whereas in the latter they are proteolytically digested and analysed as peptides.

8.4.1.2 Bottom-up proteomics

We have already seen in Chapter 6 how MS/MS can be used to provide peptide sequence information. If this capability is incorporated into a workflow able to handle complex protein mixtures, then it becomes possible to identify the many components of a proteome. Two major strategies have been developed to address this challenge (summarized in Figure 8.15). The first to find widespread use employs gel electrophoresis to separate and image proteins. Two-dimensional gel electrophoresis, in which proteins are first separated by their isoelectric

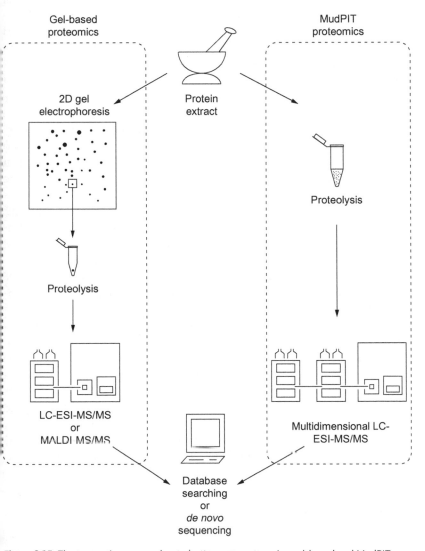

Figure 8.15 The two major approaches to bottom-up proteomics: gel-based and MudPIT proteomics.

point (pI), using a pH gradient across the gel and then by their molecular mass, by adding sodium dodecyl sulfate (SDS) to the gel, can separate mixtures of many hundreds of proteins.

Following visualization by appropriate staining, pieces of the gel containing protein 'spots' of interest can be excised and digested using a protease. This can be achieved in-gel by first shrinking the gel and swelling it with a solution containing the desired protease. Trypsin is often the protease enzyme of choice, due to its high efficiency and specificity for cleaving on the C-terminal side of Arg and Lys. The resulting peptide mixture diffuses out of the gel into the supernatant, which can then be analysed by either MALDI-MS/MS or LC-ESI-MS/MS. Peptide sequencing, and resulting protein identification, may be achieved by manual interpretation or, more commonly, database searching (see Section 8.4.1.4).

The second method used in bottom-up proteomics employs two-dimensional chromatography linked to mass spectrometry. The approach is termed 'multi-dimensional protein identification technology': MudPIT. Rather than separating proteins before digestion, in the MudPIT workflow the crude mixture of proteins is subjected to digestion and the resulting peptides are analysed (see Figure 8.15). This highly complex mixture requires high-resolution separation and high peak capacity in order to enable peptide detection and sequencing, and thereby achieve protein identification. The first chromatographic separation usually employs strong cation exchange (SCX) chromatography. A step gradient comprised of a series of discrete, increasing salt concentrations is used to elute the peptides in fractions. Each fraction is separately loaded onto the second chromatographic dimension, usually reversed-phase silica, which is linked to ESI-MS. The two-step separation approach provides sufficient separation to allow detection and sequencing of samples containing many thousands of peptides, and thus represents an extremely powerful platform for proteomic analysis.

A range of MS analyser types are used in bottom-up proteomics. The Q-TOF instrument is very popular, due to its MS/MS capability for peptide ion fragmentation and the relatively high resolving power of the TOF analyser for isotope separation of multiply-charged ions. Ion-trap spectrometers are also commonly used, but although they do not possess the resolving power of the TOF, their high scan speed and duty cycle in MS/MS mode means that they are well suited to LC-MS analysis of complex mixtures, such as those produced in proteomics studies. Providing that the peptide charge states do not exceed +3, ion traps are usually able to provide isotopic separation. FTMS, and in particular hybrid orbitrap instruments, are proving increasingly popular in proteomics. The very high resolution and mass accuracy afforded by FTMS, together with the MS/MS potential of a 'front-end' quadrupole ion trap, makes these very powerful platforms for peptide analysis and protein identification.

8.4.1.3 Top-down proteomics

In the top-down approach to proteomics, proteins are analysed intact rather than being digested into peptides. Proteins can be purified, or enriched, from complex biological matrices using affinity purification methods (targeted), or

Proteases are enzymes that digest proteins into peptides.

The term *shotgun proteomics* is often used to describe the bottom-up analysis of very complex mixtures of proteins where the objective is to capture information about as many proteins as possible. It is contrasted by *targeted proteomics*, where there is a focus on specific proteins, or groups of proteins, within a complex sample.

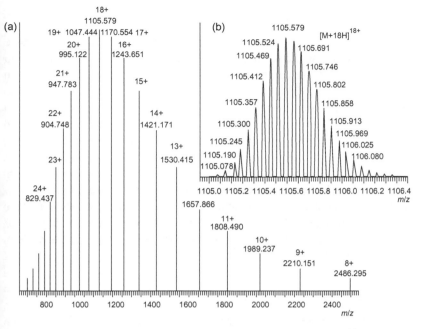

Figure 8.16 Top-down mass measurement of intact bovine beta-lactoglobulin: (a) full ESI-MS spectrum of the protein showing the range of charge states, and (b) expansion of the $[M+18H]^{18+}$ charge state showing the isotopic separation achieved at a resolution of 100,000. For sequencing, this ion can be activated and fragmented by CID, IRMPD, or ECD/ETD, or by a combination of these methods.

multi-stage chromatography (targeted or untargeted). In the latter case, separation can be performed off-line, with sample collection, or with direct coupling to the mass spectrometer. RP-LC using C4, C8, or C18 porous shell columns is preferred for MS coupling. Sequence information in top-down proteomics is provided by MS/MS on the intact protein precursor ions. Because of the high masses of proteins (typically 5–100 kDa) and their potentially large MS/MS fragments, high-resolving-power instruments are often used (Figure 8.16). FT-ICR-MS is one of the most powerful spectrometers for this purpose. Their ability to use a combination of CID, IRMPD, and ECD enables multiple rounds of orthogonal MS/MS fragmentation, which can provide maximum sequence information for intact proteins.

One of the advantages of top-down proteomics is that it has the potential to provide a complete view of any post-translational modifications (PTMs) to the protein, such as phosphorylation and glycosylation. This global view is often not possible in bottom-up analysis, where the relationship between modifications to different parts of the protein is lost as it is split into individual peptides. Bottom-up proteomics does, have the general advantage of allowing the identification of more proteins per sample, and enabling 'difficult' proteins, such as membrane proteins, to be handled by conversion to their peptides.

8.4.1.4 Database searching for protein identification

Protein identification from MS/MS data can be achieved by automated database searching, providing the sequence of a protein is known. Where an organism has had its genome sequenced, these data can be used to construct a proteome database containing the sequences of thousands of proteins. By searching the large number of MS/MS spectra resulting from a proteomic analysis, against such a database, many proteins can be identified rapidly. In theory, a sequence of 6–8 contiguous amino acids, within a peptide (bottom-up) or protein (top-down), could be sufficient to assign uniquely the identity of the protein from which it derives. Greater coverage of the protein sequence, however, provides more confidence in the identification, and this is usually required for good-quality characterization. In bottom-up approaches this is achieved by sequencing multiple peptides from each protein, whereas in top-down proteomics additional MS/MS fragment ions, together with accurate mass measurement of the protein, increases confidence.

8.4.1.5 Quantitative proteomics

Proteomics is not only concerned with the identification of proteins and post-translational modifications, but also with measuring differences in protein concentration between different sample types or treatments, or over time. A number of tools and techniques have been developed to facilitate quantification, and some of the most common methods are summarized in Table 8.3. As can be seen from the table, many approaches employ either isotopically-labelled internal standards, or isotopically-labelled chemical tags. Isotopically-labelled peptide standards permit absolute quantitation of natural peptides for which they are the isotopologues, and thus are used in targeted proteomics strategies. Isotopically-labelled tags are designed to label sets of peptides differentially. This allows samples to be combined and measured in one run whilst maintaining knowledge of peptide origin. The benefit of this approach is that it overcomes any lack of reproducibility between sample measurements. Labelled tags are useful for quantification in shotgun proteomics, as the approach does not require prior knowledge of the proteins of interest. Figure 8.17 shows the principle of isotopically-coded labelling, where relative intensities of the mass tagged peptides are measured. Other

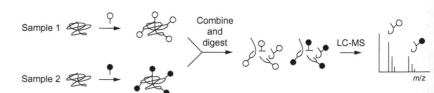

Figure 8.17 Isotope-labelled tagging strategy for quantitative proteomics. Following labelling of the protein samples with either light (○) or heavy (●) isotope tags, they are mixed, digested, and the resulting peptides analysed by LC-MS to determine the relative amounts of each protein.

Table 8.3 Comparison of the main strategies used for quantitative proteomics.

Method	Principle	Notes
Label-free quantification	Compares the relative intensities of peptide or protein precursor ions between biological samples	Does not require isotope labelling Stable instrument conditions essential Used for relative quantitation
Isotope-coded affinity tag (ICAT)	Uses an isotopically-labelled chemical tag to modify (usually) cysteine residues on peptides Samples are differentially labelled with either heavy (^{2}H or ^{13}C) or light (^{1}H or ^{12}C) isotope-coded tags, before being mixed, affinity purified, and co-analysed Relative intensities of light vs heavy labelled peptide precursor ions compared	Isotopic tag allows assignment of peptide to sample Co-analysing samples reduces uncertainty associated with separate measurements Affinity tag (biotin) allows enrichment of tagged peptides Used for relative quantitation
Isobaric tags for relative and absolute quantification (iTRAQ)	Similar concept to ICAT except tags are isobaric (each tag contains a different distribution of isotopes, but which adds up to the same total mass) Labels lysine residues and the N-terminus of protein All peptide precursor ions have the same mass, but under MS/MS the tag dissociates to produce different mass fragments to indicate origin and intensity	Unlike ICAT, tag does not contain affinity probe Isobaric nature means a single precursor ion m/z fragmented Mass difference revealed in low m/z fragment ions Used for relative and absolute quantitation
Absolute quantification (AQUA)	Uses an isotopically-labelled peptide isotopologue as an internal standard Mimics a peptide produced by proteolysis of the protein sample Sample spiked with known amount of labelled peptide and ion signals of labelled and natural peptides measured	A particular (pre-identified) peptide is targeted Used for absolute quantitation
Quantification using concatenated peptides QconCAT	Uses synthetic gene technology to produce an isotopically-labelled artificial polypeptide made-up of a series of (usually) tryptic peptides expected from the sample The sample is spiked with the labelled polypeptide and digested The resulting labelled peptides act as internal standards for those derived from the biological sample	Particular (pre-identified) peptides are targeted Used for absolute quantitation Multiple-standard peptides are produced by recombinant protein expression rather than individual chemical synthesis (as in AQUA)
Stable isotope labelling by amino acids in cell culture (SILAC)	Cell cultures are grown in media containing either unlabelled or isotopically-labelled amino acids Crude extracts from both cultures are combined, digested and the peptides analysed The signal abundance of labelled vs unlabelled peptides is compared	Used for relative quantitation A 'pulsed SILAC' variant can be used by adding amino acids at a discrete time to monitor rates of protein synthesis/turnover

strategies, such as iTRAQ (see Table 8.3), use MS/MS to generate diagnostic MS/MS product ions.

8.4.1.6 Post-translational modifications (PTMs)

Mass spectrometry is ideally suited for the detection (and quantification) of protein PTMs, as each is associated with a characteristic mass shift (or shifts). MS/MS

The process of protein synthesis in a cell is known as 'translation'. It is facilitated by the ribosome, which reads the genetic code and converts it into an amino acid sequence. Any modifications to the protein that occur after this step are known as 'post-translational modifications' (PTMs). Examples of PTMs include phosphorylation, glycosylation, and acetylation.

analysis of either peptides or intact proteins allows the site of modification to be mapped to individual amino acid residues. There are many known PTMs, and two of the most common are described in Sections 8.4.1.7 and 8.4.1.8.

8.4.1.7 Phosphoproteomics

Phosphorylation is a key regulatory modification of proteins. In eukaryotic organisms, the phosphate group is usually attached to either serine, threonine, or tyrosine residues, where it can induce a conformational change in the structure of the protein, or block protein interactions. Phosphorylation is introduced by protein kinases and removed by protein phosphatases, hence the abundance of phosphorylation within a particular protein population is closely regulated. The modification causes a mass shift of 80 Da, which is readily detectable by MS measurement of either peptides or proteins. One of the major challenges for characterizing the phosphoproteome (that portion of the proteome which is phosphorylated) is the low abundance of the modification in a protein pool. Methods to enrich phosphorylated proteins or peptides include immobilized metal affinity chromatography (IMAC) and immunoprecipitation. IMAC uses a resin impregnated with metal ions (typically Fe^{3+}) to bind and retain phosphopeptides selectively. For immunoprecipitation, antibodies are raised to the phosphorylated amino acid residue and used to bind and 'pull down' phosphoproteins and phosphopeptides.

CID-MS/MS of phosphopeptides usually results in abundant loss of the phosphate group as a neutral phosphoric acid (M−98 Da): hence the detection of this neutral loss is diagnostic of a phosphorylated peptide. The facile loss of phosphoric acid under CID conditions can make mapping the site of the modification to a particular amino acid residue challenging. The non-ergodic ion activation methods of ECD and ETD (see Sections 5.3 and 6.3) are much better suited to mapping phosphorylation, as they induce fragmentation of the peptide backbone to produce c and z ions in competition with loss of the phosphate group. This enables identification of the site of modification to individual amino acids. CID and ECD/ETD are often combined in an automated data-directed analysis, where an initial CID step is used to monitor for neutral loss of phosphoric acid. A positive response is then used to trigger MS/MS analysis by ECD/ETD.

8.4.1.8 Glycoproteomics

In higher plants and animals, many proteins (especially those secreted into the extracellular environment, or found in the cell membrane) are modified by the addition of oligosaccharides. The carbohydrate chains are often large and complex, with multiple branching. These can aid with protein stability and solubility, as well as act as interaction sites for specific receptors. Glycans are usually N-linked (to asparagine) or O-linked (to serine or threonine), with the former class being more prevalent and better understood. The study of the glycosylated proteome is known as 'glycoproteomics'. As with phosphoproteomics, mass spectrometry is a key technique, but unlike phosphorylation the mass shifts

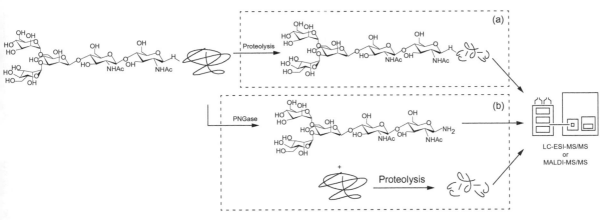

Figure 8.18 Glycoprotein analysis by mass spectrometry.

associated with glycosylation are large and varied. Glycoprotein characterization can be divided into two parts: (i) mapping the site(s) of glycosylation on the protein, and (ii) characterizing the structure of the glycan.

Mapping the sites of glycosylation within a protein can be achieved in a bottom-up approach by digesting with a protease such as trypsin and analysing the glycopeptides by MALDI-MS or (LC-)ESI-MS/MS (see Figure 8.18a). With CID MS/MS, the spectrum is often dominated by fragmentation of the glycan, whilst with ECD/ETD MS/MS fragmentation of the peptide allows the site of modification to be determined.

The most common strategy to determine the structure of an N-linked glycan is to release it from the protein using a peptide N-glycosylase (PNGase) enzyme before using MALDI-MS/MS or ESI-MS/MS (see Figure 8.18b and Section 6.3 for fragmentation modes). The protein released by PNGase treatment can be digested and analysed by LC-MS/MS. The PNGase hydrolytic action causes the asparagine linker residue to be converted to aspartic acid. Providing the sequence of the protein is known, this change, with its 1 Da mass shift, can be detected and used to map the positions of glycan attachment.

8.4.2 Native MS and structural proteomics

We have seen in the previous section, and in Chapter 6, the power of mass spectrometry for determining the primary structure of proteins (i.e. their amino acid sequence). Mass spectrometry is also able to provide information about higher protein structure and interactions. Consider the two ESI mass spectra of horse heart myoglobin shown in Figure 8.19. Spectrum (b) results from measuring the protein from a solution of 1:1 water:acetonitrile containing 0.1% formic acid. The protein displays a wide range of relatively high charge states, and gives a mass of 16,951.5 Da—the polypeptide mass of myoglobin. Spectrum (a) is obtained by electrospray ionization of the protein from a solution of ammonium acetate (25 mM) containing no organic solvent. In this spectrum the $[M+8H]^{8+}$, $[M+9H]^{9+}$, and

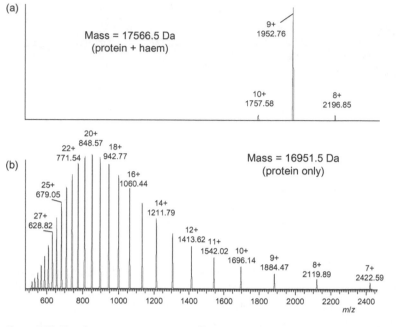

Figure 8.19 The electrospray mass spectra of horse heart myoglobin measured under (a) native and (b) denaturing conditions. The high charge states of unfolded protein are visible in (b), whilst the low charge states, characteristic of folded protein, are seen in (a). The difference in measured mass between the two spectra corresponds to the non-covalently bound haem group present in the native protein.

If a simple, routine mass measurement of a protein is required, then the sample is usually dissolved in a denaturing solvent composed of 1:1 water:acetonitrile containing 0.1% formic acid. This has a number of advantages. Most notably, it produces highly charged ions that fall within the m/z range of most mass analysers, and reduces adducts.

Myoglobin is an oxygen-binding protein found in the muscle tissue of most mammals. The protein possesses an iron-containing haem porphyrin that sits within a binding pocket, resulting in a non-covalent complex.

$[M+10H]^{10+}$ charge states dominate the spectrum, providing a mass of 17,566.5 Da. This mass corresponds to the polypeptide plus the mass of a haem porphyrin which is non-covalently bound within the protein structure in its native state. Thus, if proteins are ionized by electrospray from neutral volatile buffers, such as ammonium acetate, with carefully optimized instrumental conditions (gas pressures and voltages), it is possible to maintain non-covalent interactions. Using this approach, complexes between proteins and ligands, including cofactors, inhibitors, and drugs, can be observed and studied.

The low charge state and narrow charge distribution exhibited in Figure 8.19a is diagnostic of a compact, folded protein structure at the time when the charges were deposited in the ESI process. This is because the net average charge state of a globular protein scales with its surface area. Techniques such as ion-mobility-mass spectrometry (see Chapter 7), which are able to measure the collisional cross-section of ions, have shown that low-charge-state protein ions, in the gas phase, maintain dimensions similar to those in the condensed phase.

Mass spectrometry is not only able to measure complexes between proteins and small-molecule ligands, but also between proteins (protein–protein interactions, PPIs). Large, multiprotein complexes can be analysed, including membrane proteins and virus capsid assemblies with megaDalton (MDa) masses. This provides valuable structural information on stoichiometry, assembly, and interactions.

8.4.2.1 Protein cross-linking

As part of protein structure determination, the spatial proximity between regions of one or more protein chains can be measured by chemical cross-linking. Because of the nature of protein folding, residues that are far apart in the protein sequence (or on different polypeptide chains within a protein–protein complex) can be close in space. A bifunctional probe, with distinct spacing between two reactive sites, is used to modify the protein chain(s) at two sites (see Figure 8.20). Following proteolytic digestion, cross-linked peptides can be identified by MALDI-MS/MS or LC-ESI-MS/MS. If two residues are found to be cross-linked, then their spatial separation in the folded protein structure must have been less than the length of the spacer chain. Such information provides very useful constraints for a protein structure determination. Cross-linking probes often make use of N-hydroxysuccinimide esters, which readily react with nucleophilic lysine residues. Cross-linking efficiency is often very low, so cross-linked peptides are usually present in low abundance. To allow their enrichment, the probes can incorporate an affinity tag such as biotin. Another strategy to assist in the detection of cross-linked peptides is the use of a mixed isotope labelled probe to produce a characteristic isotopic pattern in the mass spectrum.

The non-covalent interaction between biotin, vitamin B_7, and streptavidin, a protein of bacterial origin, is one of the strongest known. It is widely exploited for affinity purification by immobilizing streptavidin on beads and attaching a biotin tag to the target molecule.

8.4.2.2 Protein hydrogen–deuterium exchange and covalent labelling

In addition to protein cross-linking, other types of protein modification can be used to gain structural information. Hydrogen–deuterium exchange (HDX) and covalent labelling are both able to report on solvent accessibility of the protein, and thereby reveal binding sites and track protein folding.

HDX works by diluting a protein sample into deuterium oxide (D_2O) to induce replacement of exchangeable protons (N-H, O-H, and S-H) with deuterium. Mass spectrometry can then be used to measure the accompanying increase in molecular mass. Regions of the protein that are readily accessible to the solvent exchange much more rapidly than those which are buried or where the hydrogens are involved in intramolecular H-bonding. In practice, the sample is

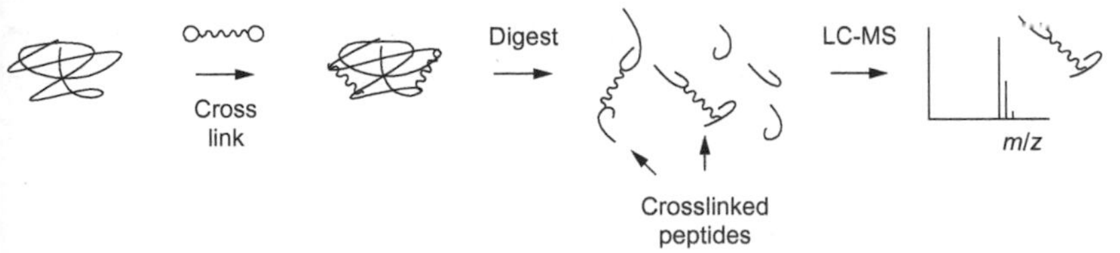

Figure 8.20 Protein cross-linking to determine special proximity between amino acid residues in a protein complex.

quenched in a low-pH solution at low temperature before measurement. This causes the very rapidly exchanging side-chain positions to exchange back to protons, whilst the relatively slow exchanging backbone peptide N-H remains unaltered under quenching conditions. As well as measuring the global HDX of an intact protein, proteolytic digestion can be used to examine exchange in specific regions. Digestion needs to be conducted under low-pH quenching conditions to prevent unwanted back exchange, and so enzymes such as pepsin that exhibit high activity under acidic conditions are employed. LC-ESI-MS analysis employing a low column temperature and an acidic solvent modifier (e.g. formic acid) helps to reduce back exchange. MS/MS of the resulting peptides has the potential to provide even higher-resolution data on solvent accessibility, but under CID conditions, hydrogen–deuterium scrambling within the peptide ion complicates interpretation. Non-ergodic MS/MS activation techniques, such as ECD and ETD, significantly reduce this problem and show considerable promise in high-resolution HDX mapping.

To overcome some of the problems associated with HDX back exchange, irreversible covalent modification of proteins has been developed. This can include the use of probes with similar reactivity to those used in protein cross-linking (e.g. acyl N-hydroxysuccinimides), but these typically exhibit a slow labelling reaction rate, which prevents the study of protein folding and transient interactions. Recently, photochemically-generated reactive intermediates such as radicals and carbenes have been employed to achieve rapid (μs to ms) labelling, and hence reveal transient species. These approaches generally require laser irradiation to deliver sufficient photon flux in a short timescale. Use of the hydroxyl radical ($^\bullet$OH) is particularly popular, due to its high reactivity and obvious similarity to the structure of water. Organic carbenes are also used, as these can interact favourably with hydrophobic patches on a protein's surface, or insert into micelles for membrane protein studies. A typical experiment would involve labelling a protein in the presence and absence of a binding partner, and then using mass spectrometry to map the sites masked by binding. This so-called 'foot-printing' approach can provide rapid, high-quality structural information using only small quantities of protein.

8.5 Mass spectrometry imaging

As well as forming mass spectra from a pure sample or complex extract, certain types of ionization also permit the acquisition of spatially discrete spectra, which allows the generation of an 'm/z image'. This can be particularly powerful for the analysis of biological material, such as tissue sections, but also for other biological and non-biological surfaces.

8.5.1 MALDI imaging

As we have seen in Chapter 2, MALDI is an important ionization method for biomolecules such as peptides and proteins. Its use of a laser to ionize a spot

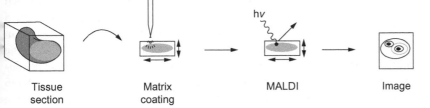

Figure 8.21 The MALDI imaging process applied to a tissue section.

on a sample target makes it highly suited for imaging applications. A schematic representation of the MALDI imaging process for tissue section analysis is shown in Figure 8.21. Following sectioning of frozen tissue using standard histology techniques, the thin section is mounted on a MALDI target and sprayed with a matrix solution. Once the solvent has evaporated, the target is introduced into the MALDI source and the laser beam rastered over the section to record a spectrum at many points across the surface. Data can be interpreted by displaying images of selected m/z values over the surface, where each pixel of the image corresponds to a particular intensity of the given ion. This is usually colour-coded to assist visualization.

A spatial resolution of <10 μm is possible with some systems, but it is essential to coat the section with a homogeneous layer of matrix to ensure optimum performance and spot-to-spot reproducibility. MALDI imaging is able to visualize both small molecules and large biomolecules such as proteins. Indeed, it is even possible to perform tryptic digestion of proteins on the surface of the section and image the resulting peptides. Among small-molecule analytes, endogenous lipids, and drugs and their metabolites, are commonly imaged, as the former can report diseased tissue, whilst the latter inform pharmacokinetic studies. It should be noted that ionization of the matrix itself can make it challenging to image some low-mass ions (~< 350 Da) due to suppression effects.

MALDI imaging can be performed by using a range of mass analysers. TOF analysers are most commonly used, but orbitrap and FT-ICR instruments have also been employed in order to perform ultra-high-resolution measurement. Spectrometers incorporating ion mobility show significant promise for imaging due to the orthogonal ion separation which they utilize. This is particularly valuable for the complex signals seen in imaging spectra.

Although MALDI imaging provides two-dimensional spatial data, it is possible to image multiple sections through tissue and use software to reconstruct a three-dimensional representation of the biological sample. Figure 8.22 illustrates the approach.

8.5.2 SIMS imaging

In principle SIMS (secondary ion mass spectrometry) imaging is analogous to MALDI imaging except that instead of using a laser to induce ionization, a beam of high-energy (keV–MeV) primary ions is rastered over the target sample. This

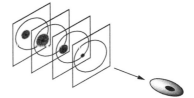

Figure 8.22 Construction of a three-dimensional tissue structure from the combination of MALDI image sections.

produces secondary ions from each spot on the sample (see Chapter 2), which can then be mass analysed. TOF analysers are commonly employed (so called TOF-SIMS), but other instrument designs are possible. The primary ion beam may consist of the metal ions Cs^+ or Ga^+, gases such as Ar^+ or O_2^+, or large molecular ions such as C_{60}^+. SIMS imaging can achieve sub-μm spatial resolution and picomole sensitivity, making it a particularly powerful method. SIMS imaging tends to be most commonly employed for analysing surfaces in materials, geological, and metallurgical studies. The high energies involved in SIMS make it a hard ionization technique, which tends to restrict it to elemental and small molecules and be less suitable for biomolecules. However, there is an increasing interest in developing SIMS approaches for the imaging of biological materials.

8.5.3 DESI imaging

As described in Chapter 2, desorption electrospray ionization (DESI) is an ambient ionization method that samples surfaces, and as such it is well suited for imaging applications. The sample to be imaged is mounted on an *x-y* microscope stage and moved with respect to a DESI probe to generate a spectrum for each region of the surface. DESI possesses lower spatial resolution (100 μm) than either MALDI or SIMS imaging, but is able to sample larger surface areas. It is capable of ionizing small molecules, peptides, and proteins effectively in an atmospheric pressure environment, which makes it a powerful imaging platform.

8.5.4 LESA imaging

Liquid extraction surface analysis (LESA) is an imaging method that uses a liquid-handling robot to form a solvent droplet contact between a pipette tip and a surface in order to extract soluble analytes. The solvent droplet is retracted and the tip moved to a chip-based nano-ESI nozzle for direct MS analysis (see Figure 8.23). This process can then be repeated over regions of the sample surface to build up an image. With a spatial resolution of 1 mm, LESA produces less well-resolved images than the other techniques mentioned previously, but it does have a number of advantages. The fact that at each point on the surface it extracts the sample means that subsequent MS analysis can be conducted over extended timescales to increase sensitivity or perform a range of MS and MS/MS analyses. In addition, unlike other imaging techniques, it is possible to link LESA to a chromatographic system prior to MS analysis, which provides potential advantages for direct quantification and in-depth coverage.

Imaging mass spectrometry is developing into a powerful approach for characterizing biological tissues and other materials. As can be seen from the previous descriptions, the choice of method depends upon the sample type/size, and the required spatial resolution and sensitivity.

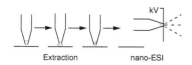

Figure 8.23 Liquid extraction surface analysis (LESA) imaging.

8.6 Summary

From the material presented in this chapter you should be familiar with the following:

- How mass spectrometry is involved in the discovery and development of new drugs.
- The way in which inborn errors in metabolism are detected using mass spectrometry.
- The role of mass spectrometry in metabolomics and its potential for i) investigating the effects of enzymes, ii) identifying biomarkers of disease, iii) elucidating metabolic pathways affected by environmental or genetic changes, and iv) the role that metabolomics can play in precision medicine.
- Using mass spectrometry for the detection and monitoring of environmental pollutants.
- The application of accelerator mass spectrometry for radiocarbon dating.
- The use of mass spectrometry in anti-doping studies for both screening and discovery.
- The application of MS in proteomics, including protein identification and quantification, as well as PTM characterization.
- The ability of native MS to detect protein non-covalent complexes.
- The use of crosslinking, HDX, and covalent labelling to study protein folding and map interactions.
- The ability of MS to provide an m/z image of surfaces or sections using MALDI, SIMS, DESI, or LESA.

8.7 Exercises

8.1 How do mass spectrometry methods for screening inborn errors in metabolism ensure quantitative results without the use of chromatography?

8.2 What are the principle differences between targeted and untargeted metabolomics?

8.3 State four analytical parameters used to increase confidence in metabolite identification in metabolomics.

8.4 Explain how thermal desorption (TD) is useful in the analysis of VOCs.

8.5 Explain why different mass spectrometry techniques might be used in anti-doping studies for i) the analysis of prohibited drugs and ii) identifying substances not yet on a prohibited list.

8.6 List three reasons why mass spectrometry is so well suited for proteomic studies.

8.7   What are the principal characteristics of top-down and bottom-up proteomics? Give one advantage of each approach over the other.

8.8   How might you enrich the phosphopeptide population in a protein digest? What is the advantage of including such a step in phosphoproteome characterization?

8.9   You suspect that three proteins A, B, and C, form a protein–protein complex. What mass spectrometry-based experiments would you perform to establish the stoichiometry of each subunit in the complex? How would you identify which protein interacts with which, and map their binding sites?

8.10 Compare MALDI, SIMS, DESI, and LESA imaging. Consider the advantages and disadvantages of each method.

8.8 Further reading

Chapman, J. R. (1993). *Practical Organic Mass Spectrometry*, 2nd edn. Chichester: Wiley.

Cole, R. B. (ed.) (2010). *Electrospray and MALDI Mass Spectrometry*. New York: Wiley.

Hoffmann, E. de and Stroobant, V. (2007). *Mass Spectrometry: Principles and Applications*, 3rd edn. Chichester: Wiley.

Schork, N. J. (2015). 'Time for one-person trials', *Nature* 520, 609–11.

Watson, J. T. and Sparkman, O. D. (2007). *Introduction to Mass Spectrometry*, 4th edn. Chichester: Wiley.

Wishart, D. S. (2016). 'Emerging applications of metabolomics in drug discovery and precision medicine', *Nature Reviews: Drug Discovery* 15, 473–84.

Glossary

Accelerating voltage (V) The voltage applied to the source to extract ions and direct them into the mass analyser.

Accelerator mass spectrometry A specialized MS method used for the detection of ^{14}C isotopes in radiocarbon dating.

Accurate mass Accurate, experimentally determined mass of an ion of known charge that can be used to determine elemental composition to within the limits defined by both the accuracy and precision of the measurement.

Ambient ionization A generic term for atmospheric pressure ionization methods where ions are formed by a source external to the mass spectrometer without sample preparation or separation. Examples include DESI and DART.

Analyser See mass analyser.

Anionated molecule Molecule converted into an anion by the addition of a negatively charged ion.

Atmospheric pressure chemical ionization (APCI) A method of ionization that uses a corona pin, held at high voltage, to generate reactive CI plasma from heated solvent vapour, which can be used to ionize dissolved sample molecules.

Atmospheric pressure ionization (API) A generic term for any method of ionization that takes place at atmospheric pressure.

Atmospheric pressure photoionization (APPI) A method of ionization that uses energy from a UV lamp to ionize sample molecules either directly or indirectly.

Atmospheric pressure solids analysis probe (ASAP) An ambient ionization method that uses a heated flow of nitrogen gas to vaporize a deposited sample and pass it over a corona pin, held at high voltage, to induce ionization.

Average mass The mass of an ion or molecule weighted for its isotopic composition.

b ions Peptide fragment ions that result from cleavage of the peptide bond and include the N-terminus.

Base peak The peak in the mass spectrum that has the greatest intensity.

Beam mass spectrometer A mass spectrometer that generates a beam of ions that pass through the instrument from source to detector (in contrast to a trapping mass spectrometer). Examples include magnetic sectors, quadrupoles, and TOF instruments.

Bottom-up sequencing A mass spectrometry-based protein sequencing method that relies on the sequencing of peptides generated by proteolytic digestion of the protein sample (in contrast to top-down sequencing).

c ions Peptide fragment ions that result from cleavage of the backbone N-Cα bond and include the N-terminus.

Cationated molecule Molecule converted into a cation by the addition of a positively charged ion.

Charge (z) The number of elementary charges e carried by an ion.

Chemical ionization (CI) A method of ionization that uses a reagent gas to ionize a sample by ion–molecule reactions.

Collision energy The energy applied to an ion through acceleration and collision with neutral gas molecules.

Collision-induced dissociation (CID) A technique of ion activation used in tandem mass spectrometry to cause the fragmentation of ions by transfer of collisional energy into internal energy.

Dalton (Da) The unit of atomic and molecular mass equal to 1/12 the mass of a ^{12}C atom.

Deconvolution The mathematical process of converting the m/z values of a series of multiply charged ions for a given molecule (usually a peptide or protein) into a single molecular mass.

Deprotonated molecule The ion $[M-H]^-$ resulting from deprotonation of a molecule. For large molecules such as peptides and proteins, multiple deprotonation may occur, especially using electrospray ionization.

Derivatization The process of converting sample molecules into a different chemical species prior to measurement. Derivatization is usually employed to improve sample stability, ionization efficiency, or chromatographic behaviour, but may also form part of a labelling strategy for sample quantitation.

Desorption electrospray ionization (DESI) A method of ambient ionization that uses the droplet spray from an electrospray probe to desorb and ionize samples from a solid surface.

Desorption ionization Any method of ionization where sample ions are generated directly by desorption from the condensed phase. Examples include MALDI and SIMS.

Detection limit See limit of detection (LOD).

Detector The part of a mass spectrometer responsible for detecting and amplifying ion current.

Direct analysis in real time (DART) A method of ambient ionization that uses a gaseous flow of metastable species, produced by a glow discharge plasma, to ionize atmospheric nitrogen and water molecules, which themselves then act as a CI gas able to ionize the sample molecules. Samples can be placed directly in the space between the probe and spectrometer API entrance orifice.

Direct insertion probe A probe equipped with a sample holder that can be used to introduce a sample, via a vacuum lock, into the source of a mass spectrometer. It is usually employed with EI, CI, and FI sources.

Double focusing mass spectrometer A sector mass spectrometer equipped with a magnetic and an electric analyser in series in such a way that ions with the same m/z but with distributions in both the direction and the translational energy of their motion are brought to a focus. It provides higher resolving power than a magnetic analyser alone.

Electron-capture dissociation (ECD) A method of ion fragmentation in tandem mass spectrometry involving the capture of low-energy electrons by multiply-protonated ions in an FT-ICR cell. The energy released by neutralization of one or more protons results in a rapid increase in internal energy of the ion, leading to fragmentation resembling that seen in ECD.

Electron ionization (EI) A method of ionization that removes one or more electrons from an atom or molecule through interactions with high-energy electrons (typically 70 eV). EI usually results in extensive fragmentation of the molecular ion M^+.

Electron-transfer dissociation (ETD) A method of ion fragmentation in tandem mass spectrometry involving the transfer of electrons from radical anions to multiply-protonated ions in an ion trap or collision cell. The energy released by neutralization of one or more protons results in a rapid increase in internal energy of the ion, leading to fragmentation resembling that seen in ECD.

Electrospray ionization (ESI) A method of atmospheric pressure ionization that uses a high voltage applied to the tip of a hollow needle containing a sample in solution. The resulting spray of charged solvent droplets leads to ionization of the dissolved sample molecules, which become desolvated by rounds of evaporation and Coulombic explosion to produce gas-phase sample ions.

Even electron ion (EE$^+$) An ion that contains no unpaired electrons in its ground state.

Exact mass The calculated mass of an ion or molecule with specified isotopic composition. Note that this is not a measured value.

Extracted ion chromatogram (EIC) A chromatogram created by plotting the intensity of the signal observed at a chosen m/z value, or set of values, in a series of mass spectra recorded as a function of retention time.

Fast atom bombardment (FAB) A method of desorption ionization that uses a stream of fast (keV) atoms (e.g. xenon) to transfer energy to a sample dissolved in a viscous liquid matrix. Ionization results from ion/molecule reactions between the ejected species.

Field free region Section of a mass spectrometer in which there are no electric or magnetic fields.

Field ionization (FI) A method of ionization by the removal of electrons from any gas-phase species *via* the action of a high electric field.

Fourier transform ion cyclotron resonance mass spectrometer (FTICR MS) A mass spectrometer that utilizes the phenomenon of ion cyclotron resonance, in combination with Fourier transformation, to measure m/z.

Fourier transform mass spectrometer (FTMS) A mass spectrometer that uses Fourier transformation of data gathered in the time domain to produce a frequency domain signal that can be converted into a mass spectrum. FTICR and orbitrap mass spectrometers both exploit this approach.

Fragment ion A product ion that results from dissociation of a precursor ion in a mass spectrometer.

Gas chromatography (GC) A method of chromatography that uses a gas as the mobile phase and an involatile liquid on a solid support as the stationary phase.

HPLC See liquid chromatography.

Hybrid mass spectrometer A mass spectrometer that possesses more than one type of mass analyser (e.g. QTOF).

Imaging mass spectrometry A procedure used to form chemically selective images of an object based on the mass spectrometric detection of ions desorbed from its surface.

Infrared multiphoton photodissociation (IRMPD) A method of ion activation/dissociation that uses IR laser radiation to transfer energy into the vibrational modes of trapped ions by the absorption of multiple photons.

Ion An atomic, molecular, or radical species with a non-zero net electric charge.

Ion cyclotron resonance (ICR) The coherent circular motion of ions moving perpendicularly to the direction of a magnetic field such that the cyclotron frequency is resonant with an applied radio frequency. The phenomenon is exploited in FTICR mass spectrometers, since cyclotron frequency is a function of an ion's m/z.

Ion mobility spectrometry (IMS) The separation of ions according to their velocity through a buffer gas under the influence of an electric field.

Ion/molecule reaction The reaction between an ion and molecule in the gas phase that usually results in either the transfer of charge, association, or fragmentation.

Ion-trap analyser A mass analyser that uses an electric field to trap ions and then eject them onto a detector in an m/z-dependent manner. Linear and three-dimensional designs are utilized.

Isobaric ions Two or more ions with the same nominal mass.

Isotope pattern A set of peaks related to ions with the same chemical formula, but containing different isotopes, that has a particular pattern associated with the relative abundance of the isotopes.

Isotope ratio mass spectrometer (IRMS) A mass spectrometer designed to measure the isotopic composition of elements with high accuracy and precision.

Isotopologue Molecules that differ only in their isotopic composition. The isotopologue of a chemical species has the same elemental composition, but at least one atom has a different number of neutrons.

Isotopomers (or isotopic isomers) Isomers with isotopic atoms. They have the same number of each isotope of each element present but differing in their positions within the molecule. The result is that the molecules are either constitutional isomers or stereoisomers solely based on isotopic location.

Lower Limit of detection (LLOD) The smallest amount of an analyte required to produce a signal that can be distinguished from background noise (usually a signal-to-noise ratio $S/N \geq 3$). Also known as lower limit of detection (LLOD).

Lower Limit of quantification (LLOQ) The smallest amount of an analyte for which a reliable relationship between concentration and signal response can be established (usually a signal-to-noise ratio $S/N \geq 10$). Also known as lower limit of quantification (LLOQ).

Liquid chromatography (LC) A method of chromatography that uses a liquid (solvent) as the mobile phase and a (usually particulate) solid as the stationary phase. High-performance liquid chromatography (HPLC) and ultra-high-performance liquid chromatography (UHPLC) are types of LC that use small particle sizes resulting in improved chromatographic efficiency and resolution.

Magnetic analyser A mass analyser that utilizes a magnetic field to deflect ions according to their m/z ratio. Also known as a magnetic sector analyser.

Mass analyser The component of a mass spectrometer that separates ions according to their m/z values.

Mass calibration The process of adjusting the mass spectrometer to measure accurate m/z values by reference to a standard calibrant, or mixture of calibrants.

Mass chromatogram See extracted ion chromatogram (EIC).

Mass defect In mass spectrometry, the difference between the nominal mass and monoisotopic mass.

Mass spectrometer (MS) An instrument for measuring the mass-to-charge ratio (m/z) of gas-phase ions.

Mass spectrum A plot of the relative abundances of ions forming a beam, or other collection, as a function of their m/z values.

Matrix-assisted laser desorption ionization (MALDI) A method of ionization producing gas-phase ions from molecules present in a solid or liquid matrix that is irradiated with a laser. The matrix is a material that absorbs the laser energy and promotes ionization.

Mean-free path of an ion The average distance an ion travels before colliding with other ions or molecules. For high-velocity ions moving through a mass spectrometer it is inversely proportional to the pressure of the gas and the collisional cross-sectional area.

Metabolomics An approach that aims to identify and quantify metabolites in complex samples in order to examine the effects of external or internal changes to a biological or environmental system. Mass spectrometry (particularly in combination with liquid chromatography) is a key technique in metabolomics studies.

Metastable ion An ion formed with internal energy higher than the threshold for dissociation but with a lifetime great enough to allow it to exit the ion source and enter the mass analyser region where it dissociates before detection.

Molecular ion An ion formed by the removal of one or more electrons from a molecule to form a positive ion (e.g. M^+), or by the addition of one or more electrons to a molecule to form a negative ion (e.g. M^-).

Monoisotopic mass The exact mass of an ion or molecule calculated using the atomic mass of the most abundant isotope of each element multiplied by the number of atoms of each element.

Multiple reaction monitoring (MRM) A tandem MS/MS method where multiple precursor-product ion pairs are monitored. It is an extension of selective reaction monitoring (SRM).

m/z the mass (m)-to-charge (z) ratio of an ion, where m is the mass in atomic mass units and z is the integer charge state.

Nano electrospray ionization (nESI) A miniaturized form of electrospray ionization that uses liquid flow rates in the nL min^{-1} range.

Nano liquid chromatography A method of chromatography that uses very low volumes resulting in nanolitre flow rates and injection volumes. Interfacing with mass spectrometry can enable higher-sensitivity analysis than when using conventional chromatographic approaches.

Neutral loss Loss of an uncharged species (usually a molecule or radical) from an ion during dissociation.

Neutral loss scan A tandem MS/MS scan that records a fixed neutral loss from precursor ions to product ions.

Nitrogen rule A rule stating that organic molecules containing any of the elements C, H, N, O, P, S, Si, and any halogens will have (i) an odd nominal molecular mass if they contain an odd number of nitrogen atoms (N_1, N_3, N_5, etc.), or (ii) an even nominal molecular mass if they contain an even number of nitrogen atoms (N_0, N_2, N_4, etc.).

Nominal mass Mass of an ion or molecule calculated using the isotopic mass of the most abundant constituent elemental isotope rounded to the nearest integer value and multiplied by the number of atoms of each element.

Odd electron ion (OE$^+$) An ion containing unpaired electrons in its ground state.

Orbitrap analyser A mass analyser composed of a pair of elongated cup-shaped outer electrodes containing a central inner electrode, or spindle. Ions injected into the evacuated space between the outer and inner electrodes orbit the spindle due to a balance of electrostatic and centrifugal forces. The oscillation frequency of ions between the two cup electrodes is a function of their m/z values and is determined by Fourier transform of time-domain data.

Post-translational modification (PTM) A modification to a protein following its translation by the ribosome. Examples include glycosylation, phosphorylation, and selective peptide bond cleavage. Mass spectrometry is often used in the characterization of such modifications.

Precursor ion An ion that reacts (usually by dissociation) to form particular product ions, or undergoes specified neutral losses.

Precursor ion scan A tandem MS/MS scan that records the precursor ions of a selected product ion or ions.

Product ion An ion formed from reaction (usually dissociation) of a particular precursor ion.

Product ion scan A tandem MS/MS scan that records the product ions of a selected precursor ion or ions.

Proteomics The study of proteomes, where a proteome is a set of proteins produced by an organism, system, or biological context. Mass spectrometry is a key tool for the detection, identification, and quantification of proteins in proteomic studies.

Protonated molecule The ion $[M+H]^+$ resulting from protonation of a molecule. For large molecules such as peptides and proteins, multiple protonation may occur, especially using electrospray ionization.

Quadrupole analyser A mass analyser that uses a combination of DC and RF voltages applied to four parallel rods in order to separate ions by m/z. Certain combinations of DC and RF voltages lead to stable trajectories for specific m/z values, with others being filtered out. A mass spectrum is generated by scanning both voltages over a defined range. By setting the DC voltage to zero, the quadrupole acts as an ion guide for focused transmission of all ions.

Reflectron-time-of-flight analyser (re-TOF) A TOF analyser incorporating a reflectron ion mirror to focus ions of the same m/z, but with distributions in translational kinetic energy (and hence distributions in flight time) onto the detector. The result is an increase in resolving power.

Relative intensity The ratio of intensity of a resolved peak to the intensity of the resolved peak that has the greatest intensity (base peak). This ratio is generally measured as the normalized ratio of the heights of the respective peaks in the mass spectrum, with the height of the base peak taken as 100. Note that relative intensity is a measure of detector response, which is not necessarily equal to relative ion abundance.

Resolution (R) The separation of m/z values in a mass spectrum given by the ratio $(m/z)/\Delta(m/z)$. $\Delta(m/z)$ may be defined as either the separation in m/z of two peak maxima of equal intensity resolved at 10% relative intensity (10% valley definition), or the full peak width at half maximum intensity (FWHM definition). It is important to specify which definition is used as, for any given case, the FWHM value of R will be twice that determined using the 10% valley definition.

Resolving power A measure of the ability of a mass analyser to provide a specified value of mass resolution.

Secondary ion mass spectrometry (SIMS) A method of ionization that uses a high-energy (keV) primary ion beam (e.g. Cs^+) to transfer energy to a sample on a solid surface and produce secondary ions from the sample.

Selected ion monitoring (SIM) Operation of a mass spectrometer in which the abundances of ions of one or more specific m/z values are recorded rather than the entire mass spectrum.

Sensitivity Slope of the calibration curve between detector response and amount of analyte.

Sodiated molecule Molecule converted into a cation by the addition of a sodium ion.

Source The region of the mass spectrometer responsible for producing sample ions.

Surface induced dissociation (SID) A method of ion activation/ dissociation that transfers energy into ions by collision with an inert surface.

Sustained off-resonance irradiation (SORI) An ion activation method performed in an FTICR MS using a radio frequency that is slightly off-resonance with the cyclotron frequency of the precursor ion. This permits sustained irradiation and activation (through collisional processes) without causing the ion orbit to exceed the dimensions of the ICR cell.

Tandem mass spectrometry (MS/MS) Acquisition and study of the spectra of the product ions or precursor ions of m/z selected ions, or of precursor ions of a selected neutral mass loss.

Time-of-flight analyser (TOF) A mass analyser that separates ions by m/z in a field free region after acceleration through a fixed accelerating voltage V. Ions of the same initial translational energy and different m/z require different times to traverse a given distance in the field free region.

Top-down sequencing A mass spectrometry-based protein sequencing method performed on intact proteins without prior digestion to peptides (in contrast to bottom-up sequencing).

Total ion current (TIC) The sum of all the separate ion currents carried by the ions of different m/z contributing to a complete mass spectrum or in a specified m/z range of a mass spectrum.

Total ion current chromatogram (TICC) The total ion current plotted against retention time.

UHPLC See liquid chromatography.

Ultraviolet photodissociation (UVPD) A method of ion activation/ dissociation that uses UV laser radiation to transfer energy into the electronic excited states of ions.

Unimolecular dissociation A fragmentation reaction in which the molecularity of the reaction system is unity. The dissociation may arise from the extra energy acquired by a metastable ion produced in the ion source or from that provided by collisional excitation of a stable ion.

y ions Peptide fragment ions that result from cleavage of the peptide bond and include the C-terminus.

z ions Peptide fragment ions that result from cleavage of the backbone N-Cα bond and include the C-terminus.

Index

Lightning Source UK Ltd.
Milton Keynes UK
UKHW051603160822
407366UK00006B/358